Marx Machiavelli Joyce
Abbott Hardy Montaigne Chesterton Austen
Defoe Melville Cooper Emerson Hugo
Haggard Eliot Grimm
Stoker Carroll Christie Molière
Wilde Maupassant Byron Schiller
Garnett Engels
Goethe Fitzgerald Hawthorne Smith Kafka
Cotton Einstein Dostoyevsky Hall
Baum Henry Kipling Doyle Willis
Dumas Nietzsche
Leslie Flaubert Turgenev Balzac
Stockton Vatsyayana Crane
Burroughs Verne
Curtis Tocqueville Vinci
Homer Widger Gogol Busch
Darwin Tolstoy Whitman
Potter Freud Thoreau Scott
Kant Zola Twain Plato Harte
Jowett Lawrence Dickens Hesse
Andersen Cervantes Burton
London Descartes Voltaire
Poe Aristotle Wells
Hale James Hastings Cooke
Bunner Shakespeare Irving
Richter Chambers
Doré da Benedict Alcott
Dante Shaw Pushkin
Swift Chekhov Wodehouse
Newton

# tredition®

tredition was established in 2006 by Sandra Latusseck and Soenke Schulz. Based in Hamburg, Germany, tredition offers publishing solutions to authors and publishing houses, combined with worldwide distribution of printed and digital book content. tredition is uniquely positioned to enable authors and publishing houses to create books on their own terms and without conventional manufacturing risks.

For more information please visit: www.tredition.com

## TREDITION CLASSICS

This book is part of the TREDITION CLASSICS series. The creators of this series are united by passion for literature and driven by the intention of making all public domain books available in printed format again - worldwide. Most TREDITION CLASSICS titles have been out of print and off the bookstore shelves for decades. At tredition we believe that a great book never goes out of style and that its value is eternal. Several mostly non-profit literature projects provide content to tredition. To support their good work, tredition donates a portion of the proceeds from each sold copy. As a reader of a TREDITION CLASSICS book, you support our mission to save many of the amazing works of world literature from oblivion. See all available books at www.tredition.com.

 Project Gutenberg

The content for this book has been graciously provided by Project Gutenberg. Project Gutenberg is a non-profit organization founded by Michael Hart in 1971 at the University of Illinois. The mission of Project Gutenberg is simple: To encourage the creation and distribution of eBooks. Project Gutenberg is the first and largest collection of public domain eBooks.

# The Genus Pinus

George Russell Shaw

# Imprint

This book is part of TREDITION CLASSICS

Author: George Russell Shaw
Cover design: Buchgut, Berlin – Germany

Publisher: tredition GmbH, Hamburg - Germany
ISBN: 978-3-8472-1905-7

www.tredition.com
www.tredition.de

Copyright:
The content of this book is sourced from the public domain.

The intention of the TREDITION CLASSICS series is to make world literature in the public domain available in printed format. Literary enthusiasts and organizations, such as Project Gutenberg, worldwide have scanned and digitally edited the original texts. tredition has subsequently formatted and redesigned the content into a modern reading layout. Therefore, we cannot guarantee the exact reproduction of the original format of a particular historic edition. Please also note that no modifications have been made to the spelling, therefore it may differ from the orthography used today.

# THE GENUS PINUS

# PUBLICATIONS OF THE ARNOLD ARBORETUM No. 5

THE
GENUS PINUS

BY

# GEORGE RUSSELL SHAW

Es giebt jedoch auch Arten—und dieses ist für den Systematiker wie für den Physiologen gleich wichtig—welche sich den wechselnden Bedingungen der Feuchtigkeit so vollkommen anpassen, dass ihre extremen Formen zu ungleichen Arten zu gehören scheinen.

*Schimper.*

CAMBRIDGE
PRINTED AT THE RIVERSIDE PRESS
1914

REPRINTED 1958 BY THE MURRAY PRINTING COMPANY
FORGE VILLAGE, MASSACHUSETTS

# INTRODUCTION

This discussion of the characters of Pinus is an attempt to determine their taxonomic significance and their utility for determining the limits of the species. A systematic arrangement follows, based on the evolution of the cone and seed from the comparatively primitive conditions that appear in Pinus cembra to the specialized cone and peculiar dissemination of Pinus radiata and its associates. This arrangement involves no radical change in existing systems. The new associations in which some of the species appear are the natural result of another point of view.

Experience with Mexican species has led me to believe that a Pine can adapt itself to various climatic conditions and can modify its growth in response to them. Variations in dimensions of leaf or cone, the number of leaves in the fascicle, the presence of pruinose branchlets, etc., which have been thought to imply specific distinctions, are often the evidence of facile adaptability. In fact such variations, in correlation with climatic variation, may argue, not for specific distinction, but for specific identity. The remarkable variation in the species may be attributed partly to this adaptability, partly to a participation, more or less pronounced, in the evolutionary processes that culminate in the serotinous Pines.

# PART I

# CHARACTERS OF THE GENUS

### THE COTYLEDON. Plate I, figs. 1-3.

The upper half of the embryo in Pinus is a cylindrical fascicle of 4 to 15 cotyledons (fig. 1). The cross-section of a cotyledon is, therefore, a triangle whose angles vary with the number composing the fascicle. Sections from fascicles of 10 and of 5 cotyledons are shown in figs. 2 and 3. Apart from this difference cotyledons are much alike. Their number varies and is indeterminate for all species, while any given number is common to so many species that the character is of no value.

### THE PRIMARY LEAF. Plate I, figs. 4-6.

Primary leaves follow the cotyledons immediately (fig. 4) and assume the usual functions of foliage for a limited period, varying from one to three years, secondary fascicles appearing here and there in their axils. With the permanent appearance of the secondary leaves the green primaries disappear and their place is taken by bud-scales, which in the spring and summer persist as scarious bracts, each subtending a fascicle of secondary leaves. At this stage the bracts present two important distinctions.

1. The bract-base is non-decurrent, like the leaf-base of Abies    fig. 5.
2. The bract-base is decurrent, like the leaf-base of Picea    fig. 6.

The two sections of the genus, Haploxylon and Diploxylon, established by Koehne on the single and double fibro-vascular bundle of the leaf, are even more accurately characterized by these two forms of bract-insertion. The difference between them, however, is most obvious on long branchlets with wide intervals between the leaf-fascicles.

The bracts of spring-shoots are the scarious bud-scales of the previous winter; but the bracts of summer-shoots have the form and green color of the primary leaf.

## THE BUD. Plate I, figs. 7-11.

The winter-bud is an aggregate of minute buds, each concealed in the axil of a primary leaf converted into a scarious, more or less fimbriate, bud-scale. Buds from which normal growth develops appear only at the nodes of the branches. On uninodal branchlets they form an apical group consisting of a terminal bud with a whorl of subterminal buds about its base. On multinodal branchlets the inner nodes bear lateral buds which may be latent.

Fig. 7 represents a magnified bud of P. resinosa, first immersed in alcohol to dissolve the resin, then deprived of its scales. This bud contains both fascicle-buds, destined for secondary leaves, and larger paler buds at its base. These last are incipient staminate flowers, sufficiently developed for recognition. Such flower-bearing buds are characteristic of the Hard Pines in distinction from the Soft Pines whose staminate flowers cannot be identified in the bud.

The want of complete data leaves the invariability of this distinction in question, but with all species that I have examined, the flowers of Hard Pines are further advanced at the end of the summer. In the following year they open earlier than those of Soft Pines in the same locality. The staminate flowers of some Hard Pines (resinosa, sylvestris, etc.,) are not apparent without removing the bud-scales, but, with most Hard Pines, they form enlargements of the bud (fig. 9).

2

Invisible or latent buds are present at the nodes and at the apex of dwarf shoots. The former are the origin of the numerous shoots that cover the trunk and branches of P. rigida, leiophylla and a few other species (fig. 10). The latter develop into shoots in the centre of a leaf-fascicle (fig. 11) when the branchlet, bearing the fascicle, has been injured.

The size, color and form of buds, the presence of resin in quantity, etc., assist in the diagnosis of species. Occasionally a peculiar bud, like that of P. palustris, may be recognized at once.

## THE BRANCHLET. Plate I, figs. 12-14.

The branchlet, as here understood, is the whole of a season's growth from a single bud, and may consist of a single internode (uninodal, fig. 12-a) or of two or more internodes (multinodal, fig. 13), each internode being defined by a leafless base and a terminal node of buds.

The spring-shoot is uninodal in all Soft Pines and in many Hard Pines, but, in P. taeda and its allies and in species with serotinous cones, it is more or less prevalently multinodal.

The uninodal spring-shoot may remain so throughout the growing season and become a uninodal branchlet. Or a summer-shoot may appear on vigorous branches of any species with the result of converting a uninodal spring-shoot into an imperfect multinodal branchlet. The summer-shoot may be recognized, during growth, by its green, not scarious bracts and, at the end of the season, by the imperfect growth of its wood and foliage (fig. 14).

The perfect multinodal branchlet is formed in the winter-bud (fig. 8-a) and the spring-shoot is multinodal. It is gradually evolved among the Hard Pines, where it may be absent, rare, frequent or prevalent, according to the species. In fact there is, in Pinus, an evolutionary tendency toward multinodal growth, with its beginnings in the summer-shoot and its culmination in the multinodal winter-bud, most prevalent among the serotinous Pines.

The multinodal shoot is never invariable in a species, but is rare, common or prevalent. This condition prevents its employment for grouping species. For Pines are not sharply divided into multinodal and uninodal species, and no exact segregation of them, based on this difference, is possible. In fact the character is unequally developed among closely related species, such as P. palustris and caribaea. Both produce multinodal shoots, but the former so rarely that it should be classed as a uninodal species, while the latter is characteristically multinodal. The multinodal spring-shoot, however, has a certain correlative value in its relation to other evolutionary processes that are obvious in the genus.

The length of the branchlet is much influenced by different soils and climates. In species able to adapt themselves to great changes, the length of the internode may vary from 50 cm. or more to 1 cm. or less. In the latter case the branch is a series of very short leafless joints terminated by a crowded penicillate tuft of leaves (fig. 12-b). Such a growth may be seen on any species (ponderosa, albicaulis, resinosa, etc.) that can survive exposure and poor nourishment.

The presence of wax, as a bloom on the branchlet, is associated with trees in arid localities, especially Mexico, where it is very common. With several species the character is inconstant, apparently dependent on environment, and is a provision against too rapid transpiration.

The branchlet furnishes evidence of the section to which the species belongs, for the bract-bases persist after the bracts have fallen away. The color of the branchlet, its lustre, the presence of minute hairs, etc., are often suggestions for determining species.

3

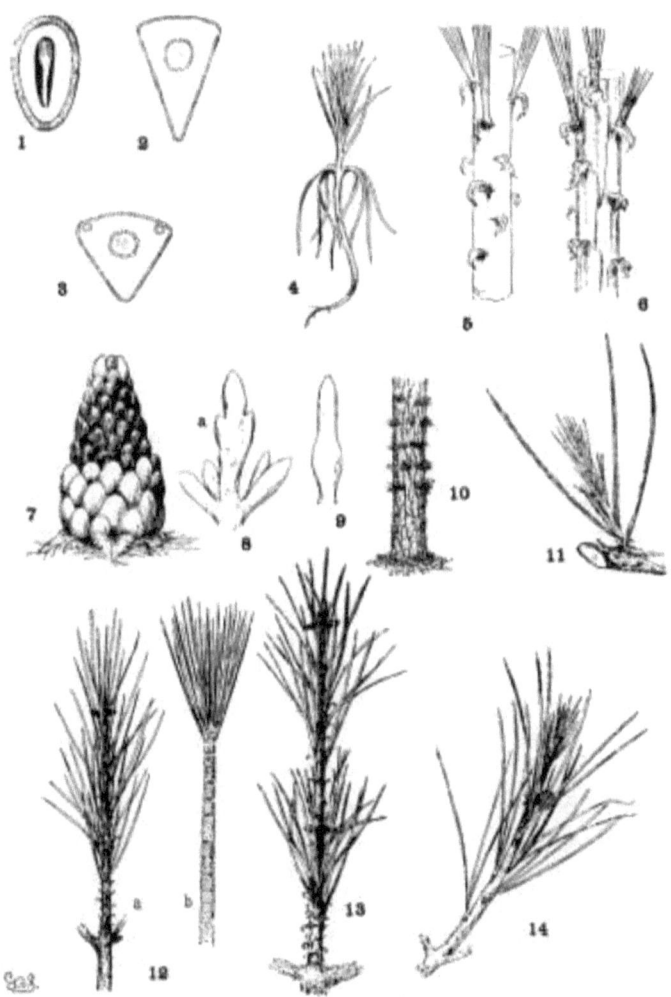

PLATE I. PRIMARY LEAF, BUD AND BRANCHLET

## THE SECONDARY LEAF. Plate II.

Secondary leaves, the permanent foliage of Pines, are borne on dwarf-shoots in the axils of primary leaves. They form cylindrical fascicles, rarely monophyllous, prevalently of 2, 3 or 5 leaves, occasionally of 4, 6, 7, or 8 leaves. The scales of the fascicle-bud elongate into a basal sheath, dec 4 iduous (fig. 15) in all Soft Pines except P. Nelsonii, persistent (fig. 16) in all Hard Pines except P. leiophylla and Lumholtzii. Inasmuch as these three species are easily recognized, the fascicle-sheath is useful for sectional distinctions.

**EXTERNAL CHARACTERS.**

The number of leaves in the fascicle is virtually constant in most species, the variations being too rare to be worthy of consideration. With some species, however, heteromerous fascicles are normal. The influences that cause this variation are not always apparent (echinata, etc.), but with P. ponderosa, leiophylla, sinensis and others, the number of leaves in the fascicle is, in some degree, dependent on climatic conditions, the smaller number occurring in colder regions. In Mexico, for example, where snow-capped mountains lie on subtropical table-lands and extremes of temperature are in juxtaposition, the conditions are favorable for the production of species with heteromerous fascicles, and the number of leaves in the fascicle possesses often climatic rather than specific significance.

Among conifers, the leaf of Pinus attains extraordinary length with great variation, from 5 cm. or less to 50 cm. or more, the maximum for each species being usually much more than twice the minimum. Climate is the predominating influence; for the shortest leaves occur on alpine and boreal species, the longest leaves on species in or near the tropics.

The length of the leaf is complicated by the peculiarities of individual trees and by pathological influences; as a general rule, however, the length of leaves is less or greater according to unfavorable or favorable conditions of temperature, moisture, soil and exposure. Therefore the dimensions of the leaf may be misleading. It can be said, however, that certain species always produce short leaves, others leaves of medium length, and others very long leaves.

Persistence of the leaf varies with the species and with the individual tree. But it is noteworthy that the longest persistence is associated with short leaves (Balfouriana, albicaulis, montana, etc.).

## INTERNAL CHARACTERS.

Since the leaf-fascicle is cylindrical, the cross-section of a leaf is a sector, its proportional part, of a circle. Theoretically the leaf, in section, should indicate the number of leaves composing its fascicle. This is absolutely true for fascicles of two leaves only. No fascicle of five leaves, that I have examined, is equally apportioned among its five members. It may be divided in various ways, one of which is shown in fig. 18, where the leaf (a) might be mistaken for one of a fascicle of 3, and the leaf (b) for one of a fascicle of 6. Therefore if absolute certainty is required, a fascicle of triquetral leaves is best determined by actual count.

The transverse section of a leaf may be conveniently divided into three distinct parts—1, the dermal tissues, epiderm, hypoderm and stomata (fig. 17-a)—2, the green tissue, containing the resin-ducts (fig. 17-b)—3, the stelar tissues, enclosed by the endoderm and containing the fibro-vascular bundle (fig. 17-c).

## THE DERMAL TISSUES OF THE LEAF.

The stomata of Pine leaves are depressed below the surface and interrupt the continuity of epiderm and hypoderm. They are wanting on the dorsal surface of the leaves of several Soft Pines, constantly in some species, irregularly in others. In Hard Pines, however, all surfaces of the leaf are stomatiferous. In several species of the Soft Pines the longitudinal lines of stomata are very conspicuous from the white bloom which modifies materially the general color of the foliage.

Under the action of hydrochloric acid the hypoderm is sharply differentiated from the epiderm by a distinct reddish tint, but without the aid of a reagent the two tissues do not always differ in appearance. The cells of epiderm and hypoderm may be so similar that they appear to form a single tissue. In most species, however, the epiderm is distinct, while the cells of the hypoderm are either uniform, 6 with equally thin or thick walls—or biform, with very thin walls in the outer row of cells and very thick walls in the inner row or rows of cells—or multiform, with cell-walls gradually thicker toward the centre of the leaf. These conditions may be tabulated as follows—

| | |
|---|---|
| Cells of epiderm and hypoderm similar | fig. 19. |
| Cells of epiderm and hypoderm distinct. | |
| Cells of hypoderm uniform, thin or thick | figs. 20, 21. |
| Cells of hypoderm biform | fig. 22. |
| Cells of hypoderm multiform | fig. 23. |

The biform hypoderm is not always obvious (clausa, Banksiana, etc.) where in some leaves there is but one row of cells. But with the examination of other leaves one or more cells of a second row will be found with very thick walls. Among Hard Pines there is no Old World species with a biform hypoderm. But there are a few American species with uniform hypoderm (resinosa, tropicalis, patula and Greggii); while, in some leaves of the few American Hard Pines with multiform hypoderm, the uniform hypoderm is a variation.

## THE GREEN TISSUE.

In this tissue are the resin-ducts, each with a border of cells, corresponding in appearance and in chemical reaction with the cells of the hypoderm and with thinner or thicker walls. With reference to the green tissue the foliar duct may be in one of four positions.

| | | |
|---|---|---|
| 1. External | against the hypoderm | fig. 24. |
| 2. Internal | against the endoderm | fig. 28. |
| 3. Medial | in the green tissue, touching neither hypoderm nor endoderm | fig. 26. |
| 4. Septal | touching both endoderm and hypoderm, forming a septum | fig. 30. |

Among the Soft Pines the external duct is invariable in the subsection Paracembra. It is also characteristic of the Strobi, where it is sometimes associated with a medial duct. In the Cembrae and the Flexiles, however, the ducts are external in some species, or medial or both in others, without regard to the affinities of these species.

Among the Hard Pines the external duct is characteristic of the Old World, there being but two American Pines with this character (resinosa and tropicalis). The internal duct is peculiar to Hard Pines of the New World, its presence in Old World species being extremely rare. The medial duct is common to species of both hemispheres, either alone or in association with ducts in other positions (figs. 25, 27). The septal duct is peculiar to a few species (oocarpa, tropicalis, and less frequently Pringlei and Merkusii). I have also seen it in a leaf of P. canariensis. The internal and septal ducts appear to be confined to the species of warm-temperate or tropical countries.

The number of resin-ducts of a single leaf may be limited to two or three (strobus, koraiensis, etc.), but in many species it is exceedingly variable and often large (pinaster, sylvestris, etc.). Eighteen or more ducts in a single leaf have been recorded. Such large numbers are peculiar to Pinus. Occasionally a single leaf, possibly the leaves of a single tree, may be without ducts, but this is never true of all the leaves of a species.

## THE STELAR TISSUES.

The walls of the endoderm are, in most species, uniform, but, with P. albicaulis and some species of western North America, the outer walls of the cells are conspicuously thickened (fig. 32). Both thin and thick walls may be found among the leaves of the group Macrocarpae and of the species longifolia.

The fibro-vascular bundle of the leaf is single in Soft Pines, double in Hard Pines. This distinction is employed by Koehne as the basis of his two sections, Haploxylon and Diploxylon. The double bundle is usually obvious even when the two parts are contiguous, but they are sometimes com 7 pletely merged into an apparently single bundle. This condition, however, is never constant in a Hard Pine, and a little investigation will discover a leaf with a true double bundle.

Some cells about the fibro-vascular bundle acquire thick walls with the appearance and chemical reaction of the hypoderm cells. Among the Soft Pines this condition is most obvious in the group Cembroides. Among the Hard Pines it appears in all degrees of development, being absent (figs. 24, 25), sometimes in irregular lines above and below the bundle (figs. 26, 27, 30, 31), or forming a conspicuous tissue between and partly enclosing the two parts of the bundle (figs. 28, 29).

The leaf-section furnishes sectional and other lesser distinctions. It is often decisive in separating species otherwise difficult to distinguish (nigra and resinosa or Thunbergii and sinensis, etc.). Sometimes it is sufficiently distinct to determine a species without recourse to other characters (tropicalis, oocarpa, Merkusii, etc.). An intimate knowledge of the leaf-section, with an understanding of the limits of its variation, is a valuable equipment for recognizing species.

5

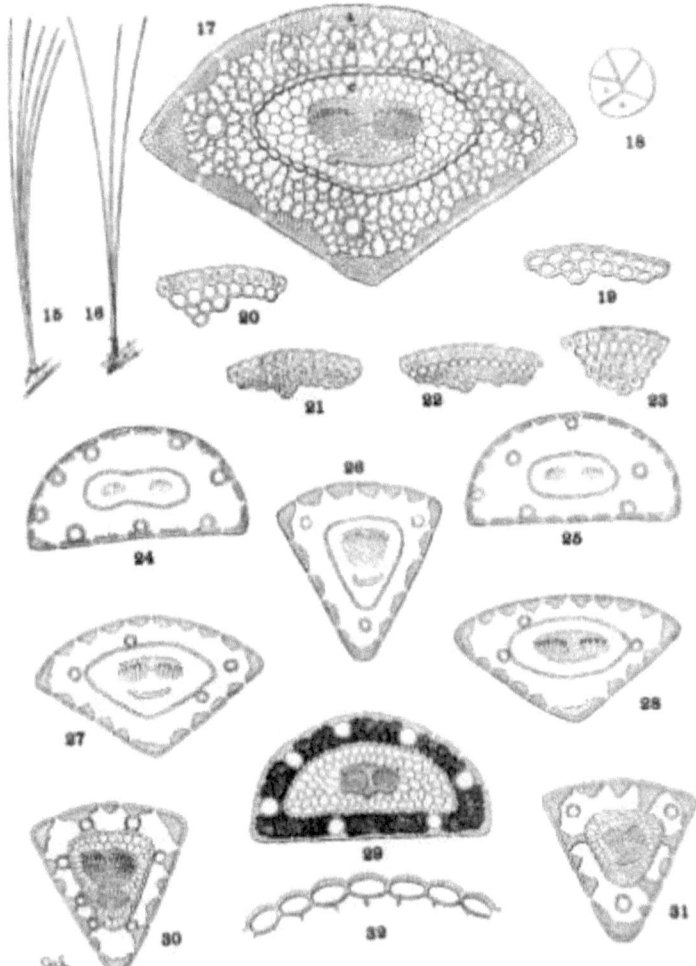

PLATE II. SECONDARY LEAVES

## THE FLOWERS. Plate III, figs. 33-39.

The flowers in Pinus are monoecious, the pistillate in the position of a long shoot, taking the place of a subterminal or lateral bud, the staminate in the position of a dwarf-shoot, taking the place of a leaf-fascicle but confined to the basal part of the internode.

Pistillate flowers are single or verticillate. On multinodal shoots they are often multiserial, appearing on two or more nodes of the same spring-shoot (fig. 33). On uninodal shoots they are necessarily subterminal (fig. 34), the lateral pistillate flower being possible only on multinodal shoots (fig. 35) where it is often associated with the subterminal flower (fig. 33). Like the multinodal shoot, on which its existence depends, the lateral pistillate flower cannot be employed for grouping the species. It is merely the frequent, but not the essential, evidence of condition of growth that is more perfectly characterized by the shoot itself.

Staminate catkins are in crowded clusters, capitate or elongate (figs. 36, 37), but with much variation in the number of catkins in each cluster. In P. rigida I have found single catkins or clusters of all numbers from two to seventy or more. In P. Massoniana and P. densiflora a cluster attains such unusual length (fig. 37) that this character becomes a valuable distinction between these species and P. sinensis, which has short-capitate clusters. The catkins differ much in size, the largest being found among the Hard Pines.

In the connective of the binate pollen-sacs there is a notable difference (figs. 38, 39), the smaller form being characteristic of the Soft Pines. But this is not invariable (excelsa, sylvestris, etc.), and the absence of complete data does not permit an accurate estimate of its importance.

## THE CONELET. Plate III, figs. 40-45.

After pollination the pistillate flower closes and becomes the conelet, the staminate flowers withering and falling away. The conelet makes no appreciable growth until the following year. Like the pistillate flower it may be subterminal or lateral, but a subterminal pistillate flower may become a pseudolateral conelet by reason of a summer-growth (fig. 40-a). Such a condition may be recognized on the branchlets of the present, and of the previous year (fig. 40-b), by the very short internode and short leaves beyond the fruit.

The conelet offers some distinctions of form, of color, and of length of peduncle, while in some species (sylvestris, caribaea, etc.) its reflexed position is an important specific character. The most important distinctions, however, are found in its scales, which may be

| | | |
|---|---|---|
| 1. entire | subsection Cembra | fig. 41. |
| 2. tuberculate | tropicalis, etc. | fig. 42. |
| 3. short-mucronate | sylvestris, glabra, etc. | fig. 43. |
| 4. long-mucronate | aristata, contorta, etc. | fig. 44. |
| 5. spinescent | taeda, pungens, etc. | fig. 45. |

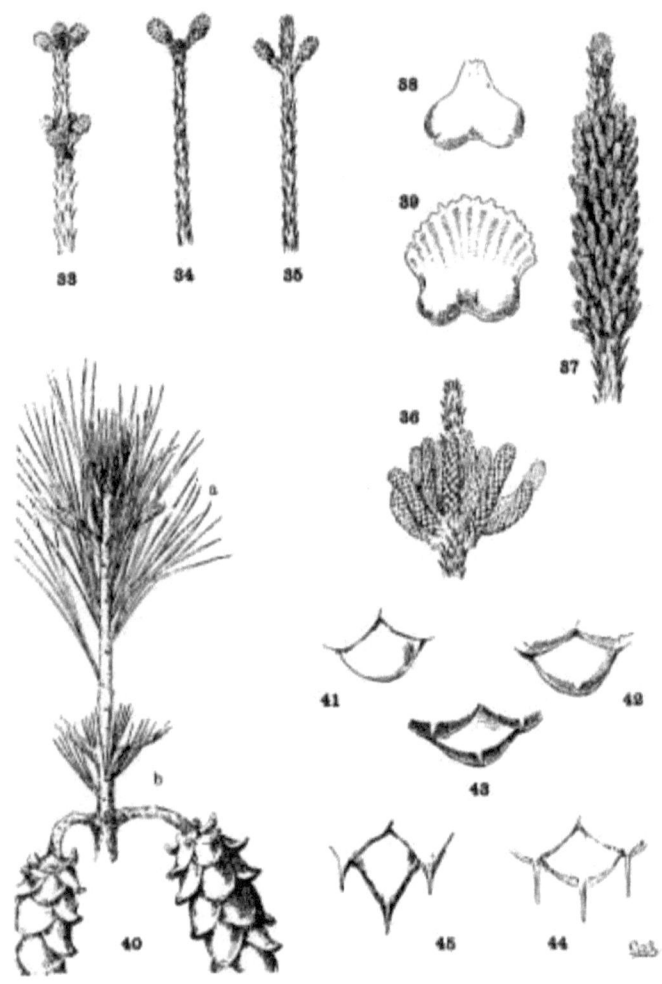

PLATE III. FLOWERS AND CONELET

## THE CONE. Plate IV.

The cone of Pinus shows great differences of color, form and tissue; these are useful for specific and sectional distinctions, while the gradual change from the primitive conditions of the Cembrae to the elaborate form, structure and mode of dissemination of some serotinous species are obvious evidence of an evolution among the species of remarkable taxonomic range. A form new among Coniferae appears, the oblique cone, and a new condition, the serotinous cone, both appearing at first alone and, finally, in constant association.

## COLOR OF THE CONE.

With few exceptions the color of the ripe cone may be classified under one of the following shades of brown or yellow.

| | |
|---|---|
| Nut-brown | The stain of the walnut-husk. |
| Rufous brown | A pronounced reddish nut-brown. |
| Fulvous brown | A yellowish nut-brown. |
| Tawny yellow | The color of the lion. |
| Orange | Ochre-yellow to red-orange. |

These colors may be paler or deeper. They may be obscured by a fuscous shade or may be modified by a dull or lustrous surface. The presence of two or more of these shades in a single species and the inherent difficulties of color description lessen the value of the character. Nevertheless certain allied species, such as P. nigra and Thunbergii, or P. densiflora and Massoniana, may be distinguished by the prevalent difference in the color of their cones.

## DIMENSIONS OF THE CONE.

The cone is small, medium or large in different species, but varies greatly under the influences of environment or of individual peculiarities. The character possesses relative value only, for great variation is possible in the same locality and even on the same tree.

## THE PEDUNCLE.

All conelets are pedunculate, but in some species the peduncle, even when long (patula), may become overgrown and concealed by the basal scales of the ripe cone. Articulation usually takes place between the peduncle and the branch, sometimes with the loss of a few basal scales which remain temporarily on the tree (ponderosa, palustris, etc.). With P. Nelsonii, and to a less degree with P. Armandi, there is articulation between the cone and its peduncle.

There are several species bearing persistent cones with no articulation. This condition appears in other genera, such as Larix and Picea, but without obvious significance. In Pinus, however, the gradual appearance of the persistent cone, for it is rare, common, prevalent or invariable in different species, and its essential association with the serotinous cone, suggest an evolution toward a definite end.

## THE UMBO.

The exposed part of the scale of the conelet is the umbo of the ripe cone, a small definite area representing the earlier part of the biennial growth of the cone. The position of the umbo on the apophysis is the basis of Koehne's subdivision of the section Haploxylon.

| | | |
|---|---|---|
| 1. Umbo terminal | Subsection Cembra | fig. 46-a. |
| 2. Umbo dorsal | Subsection Paracembra | fig. 46-b. |

Two other characters assist in establishing these subsections — the conelet, unarmed in Cembra, armed in Paracembra — the pits of the ray-cells of the wood, large in Cembra, small in Paracembra.

10

## THE APOPHYSIS.

The apophysis represents the later and larger growth of the cone-scale. With a terminal umbo the margin of the apophysis is free and may be rounded (fig. 49) or may taper to a blunt point (fig. 52), and any extension of the scale is a terminal extension. With the dorsal umbo all sides of the apophysis are confined between other apophyses, and any extension is a dorsal thickening of the apophysis or a dorsal protuberance. The outline of an apophysis with a dorsal umbo is quadrangular, or it is irregularly pentagonal or hexagonal, the different forms depending on the arrangement of the contiguous scales, whether of definite or indefinite phyllotactic order, a distinction to be considered later.

The two positions of the umbo result from the relative growth of the dorsal and ventral surfaces of the cone-scale. With the terminal umbo the growth of both surfaces is uniform, with the dorsal umbo the growth is unequal. A true terminal umbo rests on the surface of the underlying scale, although several species with terminal umbos show the first stages of the dorsal umbo. The umbo of P. Lambertiana or of P. flexilis does not touch the surface of the scale below, and a small portion of the under side of the apophysis is brought into view on the closed cone. The cone of P. albicaulis (Plate VIII, fig. 90) shows all degrees of development between a terminal umbo near the apex of the cone and a dorsal umbo near its base.

The growth of the apophysis may be limited and constant (strobus, echinata, etc.) or exceedingly variable, ranging from a slight thickness to a long protuberance (pseudostrobus, montana, etc.). The protuberance is usually reflexed from the unequal growth of the two surfaces. With the terminal umbo the protuberance lengthens the scale, with the dorsal umbo it thickens the scale. It is sometimes a specific character (ayacahuite, longifolia) appearing on all cones of the species, sometimes a varietal form, associated in the same species with an unprolonged apophysis (sylvestris, montana).

On different parts of the same cone, base, centre or apex, the dimensions of the apophyses differ, but at each level the scales may be uniform on all sides of the cone. That is to say, the cone is symmetrical with reference to any plane passing through its axis. This, the symmetrical cone, is characteristic of all other genera of the Abietineae, and is invariable among the Soft Pines and in many Hard Pines (figs. 47, 48, 52, 54). But among the Hard Pines there is gradually developed a new form of cone with smaller flatter apophyses on the anterior, and larger thicker apophyses on the posterior surface. This is the peculiar oblique cone of Pinus (figs. 50, 51, 53), symmetrical with reference to one plane only, which includes the axis of both cone and branch. The oblique cone is a gradual development among the Hard Pines; in some species it is associated as a varietal form with the symmetrical cone, and finally, in some serotinous species, it is the constant form.

## THE OBLIQUE CONE.

When the oblique cone is merely a varietal form (halepensis, etc.), it gives the impression of an accident, resulting from the reflexed position of the cone and the consequent greater development of the scales receiving a greater amount of light and air. But with the serotinous cones (radiata, attenuata), the advantages of this form become apparent. The cones of these species are in crowded nodal clusters, reflexed against the branch (fig. 50). The inner, anterior scales are perfectly protected by their position, while the outer, posterior scales are exposed to the weather. These last only are very thick; that is to say, there is an economical distribution of protective tissue, with the greatest amount where it is most needed. The oblique form is peculiarly adapted for a cone destined to remain on the tree for twenty years or more and to preserve its seeds unimpaired. Like the persistent cone, the oblique cone finds in association with the serotinous cone a definite reason for existence.

11

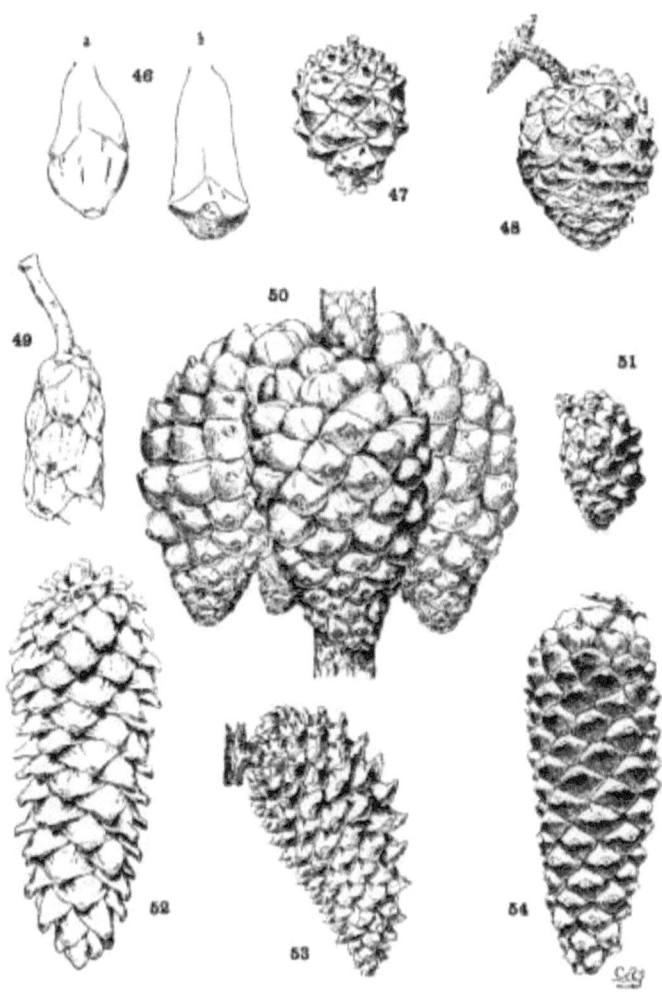

PLATE IV. THE CONE

## PHYLLOTAXIS. Plate V.

There is an obvious difference between the cones of the two sections of the genus. Those of the Soft Pines (figs. 55, 56) have larger and fewer scales, those of the Hard Pines (figs. 57, 58) have more numerous and smaller scales, in proportion to the size of the cone. The former condition represents a lower, the latter condition represents a higher, order of phyllotaxis.

## DEFINITE PHYLLOTAXIS.

On a cylindrical axis with scales of the same size, the spiral arrangement would appear as in fig. 62, where the scales are quadrangular and any four adjacent scales are in mutual contact at their sides or angles. These four scales lie on four obvious secondary spirals (fig. 59, a-a, b-b, c-c, d-d). According to the phyllotactic order of the scales these may be the spirals of 2, 3, 5, 8 or of 3, 5, 8, 13 or of 5, 8, 13, 21 etc., etc., from which combinations the primary spiral, on which the scales are inserted on the cone-axis, can be easily deduced. Four quadrangular scales in mutual contact represent the condition of definite phyllotaxis. If the cone is conical, definite phyllotaxis would be possible among all the scales only when the size of the scales diminishes in equal measure with the gradual diminution of the cone's diameter. Such a hypothetical cone is shown in fig. 61.

## INDEFINITE PHYLLOTAXIS.

On an imaginary cone of conical form and with scales of equal size throughout, there must be more scales about the base than about the apex of the cone. The phyllotactic conditions must differ, and the obvious spirals, in passing from base to apex, must undergo readjustment. If the scales at the base are in definite phyllotactic order and those at the apex are in the next lower order, it is evident that intermediate scales, in the gradual change from one condition to the other, must represent different conditions of indefinite phyllotaxis, while those in a central position on the cone may belong equally to either of two orders.

A Pine cone is never absolutely cylindrical nor do its scales vary in size proportionately to the change of diameter. Most of the scales of a cone are in indefinite phyllotactic relation, while definite phyllotaxis is found only at points on the cone.

As an extreme illustration, the cone of P. pinaster (fig. 60) shows four mutually contiguous quadrangular apophyses at (a), lying on the obvious spirals 5, 8, 13, 21, at (b) four similar apophyses on the spirals 3, 5, 8, 13, and at (c) four others on the spirals 2, 3, 5, 8. Between these three points are apophyses of irregular pentagonal or hexagonal outline, with three scales only in mutual contact (figs. 63, 64). Such are the majority of the scales of the cone and represent more or less indefinite conditions of phyllotaxis.

The cones of Hard Pines, by reason of relatively more and smaller scales and of a more conical form, attain a higher phyllotaxis and a more complex condition, two or even three orders being represented on a single cone; while the cones of Soft Pines, by reason of relatively fewer and larger scales and a more cylindrical form, are of lower phyllotaxis, with one order only more or less definitely presented. Therefore phyllotaxis furnishes another distinction between the two sections of the genus, but its further employment is exceedingly restricted on account of the constant repetition of the same orders among the species.

## THE CONE-TISSUES. Plate VI.

The axis of the cone is a woody shell, enclosing a wide pith and covered by a thick cortex traversed by resin-ducts. By removing the scales and cortex from the axis (fig. 65) the wood is seen to be in sinuous strands uniting above and below fusiform openings, the points of insertion of the cone-scales. From the wood, at each insertion, three stout strands enter the scale, dividing and subdividing 14 into smaller tapering strands whose delicate tips converge toward the umbo. Fig. 70 represents a magnified cross-section of half the cone-scale of P. Greggii; at (a) is a compact dorsal plate of bast cells; at (e) is a ventral plate of the same tissue but of less amount; at (b) is the softer brown tissue enclosing the wood-strands (d, d) (the last much more magnified in fig. 69) and the resin-ducts (e, e).

### WOOD STRANDS.

The wood-strands, forming the axis of the cone, differ in tenacity in the two sections of the genus. Those of the Soft Pines are easily pulled apart by the fingers, those of the Hard Pines are tougher in various degrees and cannot be torn apart without the aid of a tool. This difference is correlated with differences in other tissues, all of them combining in a gradual change from a cone of soft yielding texture to one of great hardness and durability.

If a cone scale of P. ayacahuite is stripped of its brown and bast tissues (fig. 66) and is immersed in water and subsequently dried, there is at first a flexion toward the cone-axis (fig. 67) and then away from it (fig. 68). The wood-strands are hygroscopic and coöperate with the bast tissues in opening and closing the cone. This appears to be true of all species excepting the three species of the Cembrae, whose strands are so small and weak that they are not obviously affected by hygrometric changes.

## BAST TISSUE.

With the exception of the three species of the Cembrae the inner part of the cone-scales is protected by sclerenchymatous cells forming hard dorsal and ventral plates (fig. 70, a, c). In Soft Pines these cells are subordinate to the more numerous parenchymatous cells, but in Hard Pines the sclerenchyma increases in amount until, among the serotinous species, it is the predominating tissue of the cone-scale, giving to these cones their remarkable strength and durability.

This bast tissue is hygroscopic and, with its greater thickness on the dorsal surface, there is a much greater strain on that side of the scale, tending to force the scales apart when they are ripe and dry, and subsequently closing and opening the cone on rainy and sunny days.

The cone, during the second season's growth, is completely closed, its scales adhering together with more or less tenacity. In most species the hygroscopic energy of the scales is sufficient to open the cone under the dry condition of its maturity, but with several species the adhesion is so persistent that some of the cones remain closed for many years. These are the peculiar serotinous cones of the genus.

## THE SEROTINOUS CONE.

As an illustration of the area to which the adhesion is confined, a section may be sawed from a cone of P. attenuata (fig. 71). The axis and the scales that have been severed from their apophyses (b) can be easily pushed out of the annulus (a), which is composed wholly of apophyses so firmly adherent that they will successfully resist a strong effort to break them apart. When immersed in boiling water, however, the ring falls to pieces. An examination of these pieces discovers adhesion only on a narrow ventral border under the apophysis and on a corresponding dorsal border back of the apophysis. The rest of the scale is not adherent, so that the seed is free to fall at the opening of the cone.

The serotinous cone is a gradual development, wanting in most species, rare in a few, less or more frequent in others. A similar evolution of the persistent cone, of the oblique cone and of the cone-tissues has been already discussed. All these progressive characters culminate in mutual association in P. radiata and its allies. The result is a highly specialized fruit that should convey taxonomic significance of some kind.

With all serotinous species that I have seen, some of the trees open their cones at maturity, others at indefinite intervals. That is to say, the seed of a prolific year is not at the mercy of a single, 16 perhaps unfavorable season. The chances of successful germination are much increased by the intermittent seed-release peculiar to these Pines. Such a method of dissemination must accrue to the advantage of a species. In other words, this intermittent dissemination and the oblique form of cone with its perfected tissues all mark the highest development of the genus.

13

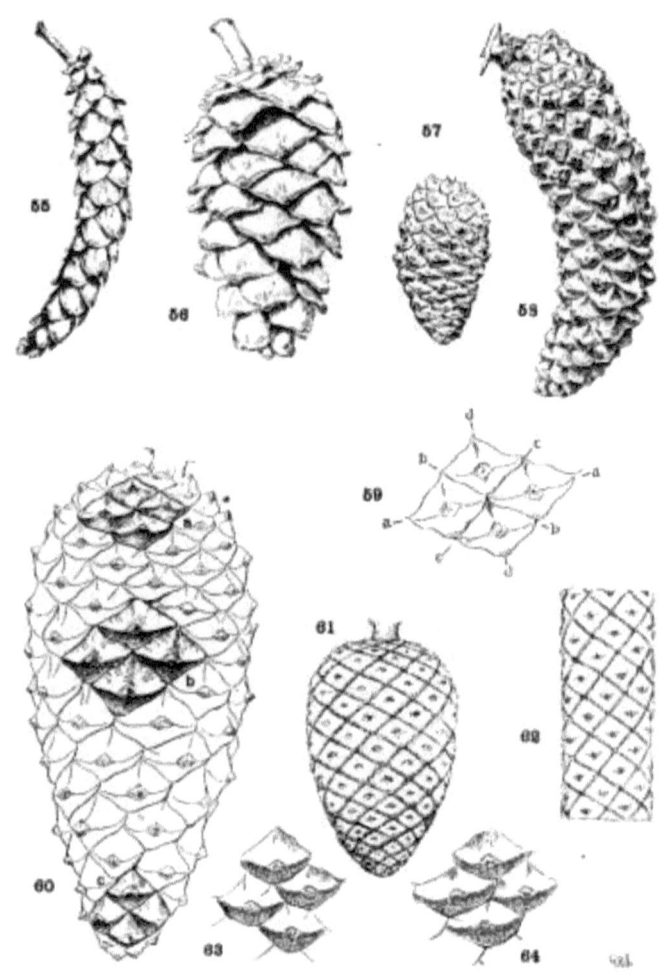

PLATE V. PHYLLOTAXIS OF THE CONE

# THE SEED. Plate VI. Figs. 72-79.

The seed of Pinus contains an embryo, with the cotyledons clearly defined, embedded in albumen, which is protected by a bony testa with an external membranous spermoderm, produced, in most species, into an effective wing. While the seed of other genera of the Abietineae shows no striking difference among the species, that of Pinus is remarkably variable, presenting alike the most primitive and the most elaborate forms among the Conifers. These differences are valuable for the segregation of kindred species and for some specific distinctions.

## WINGLESS SEEDS.

With wingless seeds the main distinction is found in the spermoderm, which is entire in one species only, P. koraiensis. In P. cembra it is wanting on the ventral surface of the nut, but on the dorsal surface, it is adnate partly to the nut, partly to the cone-scale. The nut of P. albicaulis and that of P. cembroides are quite bare of membranous cover. The spermoderm of P. flexilis is reduced to a marginal border, slightly produced into a rudimentary wing adnate to the nut.

## THE ADNATE WING.

In P. strobus, longifolia and their allies and in P. Balfouriana the spermoderm is prolonged into an effective wing-blade from a marginal adnate base like that of P. flexilis. This adnate wing cannot be detached without injury.

## THE ARTICULATE WING.

The articulate wing can be removed from the nut and can be replaced without injury. An ineffective form of this wing is seen in the Gerardianae and in P. pinea, where the blade is very short and the base has no effective grasp on the nut.

The base of the effective articulate wing contains hygroscopic tissue which acts with the hygroscopic tissue of the cone-scales. The dry conditions that open the cone and release the seeds cause the bifurcate base of the wing to grasp the nut more firmly.

This articulate wing is found in P. aristata and in all Hard Pines except P. pinea, longifolia and canariensis. The wing-blade is usually membranous throughout, but in some species there is a thickening of the base of the blade that meets the membranous apical part in an oblique line along which the wing is easily broken apart. This last condition attains in P. Coulteri and its associates a remarkable development.

Plate VI, fig. 72 shows the wingless seed of P. cembroides; fig. 73 represents the seed of P. flexilis, with a rudimentary wing; fig. 74 shows two seeds of P. strobus, intact and with the wing broken away; fig. 75 represents the articulate wing, whose bifurcate base when wet (fig. 76) tends to open and release the nut. When dry (fig. 77) the forks of the base, in the absence of the nut, close together and cross their tips; figs. 78, 79 show the peculiar reinforced articulate wing of P. Coulteri.

Such wide variation in so important an organ suggests generic difference. But here we are met by the association of the different forms in species evidently closely allied. The two Foxtail Pines are so similar in most characters that they have been considered, with good reason, to be specifically identical; yet the seed-wing of P. Balfouriana is adnate, that of P. aristata articulate. P. Ayacahuite produces not only the characteristic wing of the Strobi, adnate, long and effective, but also, in the northern variety, a seed with a rudimentary wing, the exact counterpart of the seed of P. flexilis. 17 In both sections of the genus are found the effective adnate wing (Strobi and Longifoliae) and the inefficient articulate wing (Gerardianae and Pineae). A little examination of all forms of the seed will show that they blend gradually one into another.

The color of the wing is occasionally peculiar, as in the group Longifoliae. There is usually no constancy in this character, for the wing may be uniform in color or variously striated in seeds of the same species. The length and breadth of the seed-wing, being dependent on the varying sizes of the cone-scale, differ in the same cone. They are also inconstant in different cones of the same species, and of this inconstancy the seed of P. ayacahuite furnishes the most notable example.

15

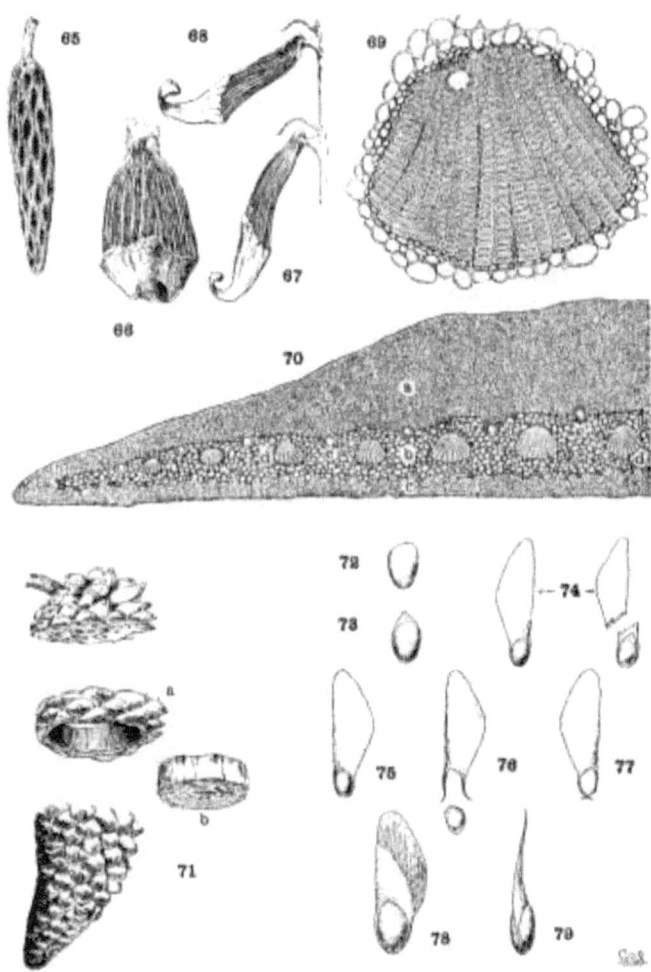

PLATE VI. CONE-TISSUES AND SEEDS

## THE WOOD. Plate VII.

With the exception of the medullary rays, a very small proportion of the whole, the wood of Pinus, as seen in cross-section (fig. 82), is a homogeneous tissue of wood-tracheids with interspersed resin-ducts. In tangential section the medullary rays appear in two forms, linear, without a resin-duct, and fusiform, with a central resin-duct. In radial section the cells of the linear rays are of two kinds, ray-tracheids, forming the upper and lower limits of the ray, characterized by small bordered pits, and ray-cells, between the tracheids, characterized by simple pits.

The walls of the ray-tracheids may be smooth or dentate; the pits of the ray-cells may be large or small. These conditions admit of four combinations, all of which appear in the medullary rays of Pinus, and of which a schematic representation is given in Plate VII. These combinations are

Ray-tracheids with smooth walls. Soft Pines.
| Ray-cells with large pits | Subsection Cembra | fig. 80. |
| Ray-cells with small pits | Subsection Paracembra | fig. 81. |

Ray-tracheids with dentate walls. Hard Pines.
| Ray-cells with large pits | Group Lariciones | fig. 83. |
| Ray-cells with small pits | Other Hard Pines | fig. 84. |

This, the simplest classification of Pine-wood, is not without exceptions. P. pinea of the Hard Pines resembles, in its wood-characters, P. Gerardiana and P. Bungeana of the Soft Pines. The dentate ray-tracheids of P. longifolia are not always obvious. The tracheids of P. luchuensis, according to Bergerstein (Wiesner Festschr. 112), have smooth walls. My specimen shows dentate tracheids. There is also evidence of transition from small to large pits (I. W. Bailey in Am. Nat. xliv. 292). Both large and small pits appear in my specimen of P. Merkusii.

Of other wood-characters, the presence or absence of tangential pits in the tracheids of the late wood establishes a distinction between Soft and Hard Pines. These pits, however, while always present in Soft Pines, are not always absent in Hard Pines. The single and multiple rows of resin-ducts in the wood of the first year may prove to be a reliable sectional distinction, but this character has not been sufficiently investigated to test its constancy. The wood-characters, therefore, however decisive they may be for establishing the phylogenetic relations of different genera, must be employed in the classification of the Pines with the same reservations that apply to external characters.

Ray-tracheids with dentate walls and ray-cells with large pits are peculiar to Pinus. Therefore the presence of these characters, alone or in combination, is sufficient evidence for the recognition of Pinewood. But the combination of smooth tracheids with small pits (subsection Paracembra) Pinus shares with Picea, Larix and Pseudotsuga.

Among Hard Pines the size of the pits has a certain geographical significance. The large pits are found in all species of the Old World except P. halepensis and P. pinaster; the small pits in all species of the New World except P. resinosa and P. tropicalis. The Asiatic P. Merkusii with both large and small pits is not strictly an exception to this geographical distinction. The four exceptional species by this and by other characters unite the Hard Pines of the two hemispheres.

18

## THE BARK.

Bark is the outer part of the cortex that has perished, having been cut off from nourishment by the thin hard plates of the bark-scales. In the late and early bark-formation is found a general but by no means an exact distinction between Soft and Hard Pines. In the Soft Pines the cortex remains alive for many years, adjusting itself by growth to the increasing thickness of the wood. The trunks of young trees remain smooth and without rifts. In the Hard Pines the bark-formation begins early and the trunks of young trees are covered with a scaly or rifted bark. The smooth upper trunk of older trees is invariable in Soft Pines, but in Hard Pines there are several exceptions to early bark-formation. These exceptions are easily recognized in the field, and the character is of decisive specific importance (glabra, halepensis, etc.).

Among species with early bark-formation are two forms of bark: 1, cumulative, sufficiently persistent to acquire thickness and the familiar dark gray and fuscous-brown shades of bark long exposed to the weather; 2, deciduous, constantly falling away in thin scales and exposing fresh red inner surfaces. The latter are commonly known as Red Pines, as distinguished from Black Pines with dark cumulative bark. Deciduous bark changes after some years to cumulative bark, and the upper trunk only of mature trees is red. Red Pines, although usually recognizable by their bark, are by no means constant in this character. Oecological or pathological influences may check the fall of the bark-scales, and then the distinction between the upper and lower parts of the trunk becomes lost.

19

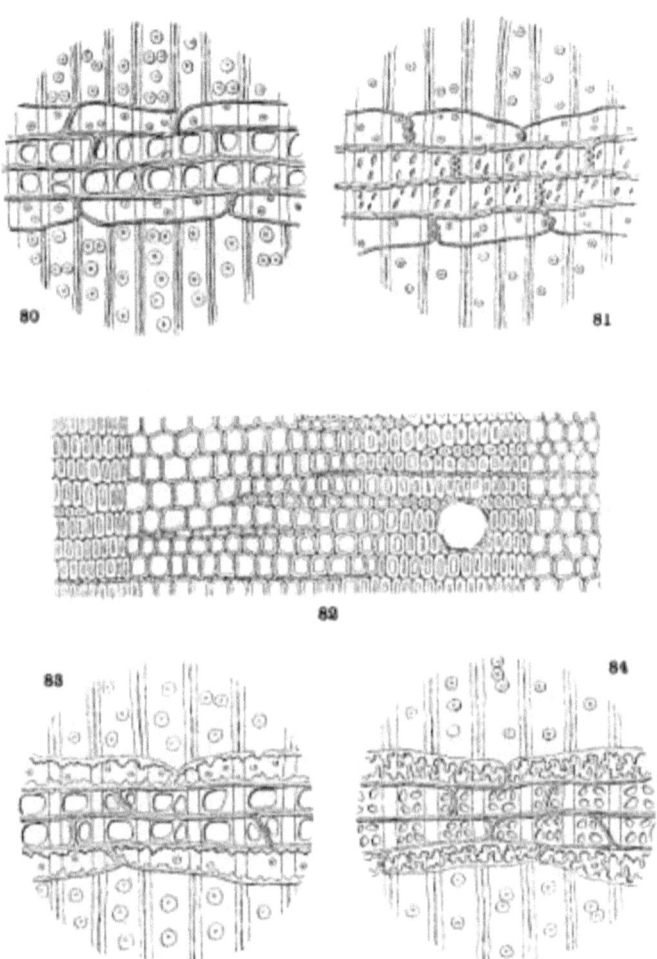

PLATE VII. THE WOOD

# SUMMARY

The various characters that have been considered in the previous pages may be classified under different heads, some of them applicable to the whole genus, others to larger or smaller groups of species.

## GENERIC CHARACTERS

Several characters, quite distinct from those of other genera, are common to all the species.

1. The primary leaf — appearing as a scale or bract throughout the life of the tree.
2. The bud — its constant position at the nodes.
3. The internode — its three distinct divisions.
4. The secondary leaves — in cylindrical fascicles with a basal sheath.
5. The pistillate flower — its constant nodal position and its verticillate clusters.
6. The staminate flower — its constant basal position on the internode and its compact clusters.
7. The cone — its clearly defined annual growths.

Pinus is also peculiar in the dimorphism of shoots and leaves and in their constant interrelations with the diclinous flowers. Evolutionary processes develop features peculiar to Pinus alone (the oblique cone, etc.), but confined to a limited number of species.

## SECTIONAL CHARACTERS

There are several characters that actually or potentially divide the genus into two distinct sections, popularly known as Soft and Hard Pines.

- 1. The fibro-vascular bundle of the leaf, single or double.
- 2. The base of the bract subtending the leaf-fascicle, non-decurrent or decurrent
- 3. The phyllotaxis of the cone, simple or complex.
- 4. The flower-bud, its less or greater development.

Some characters indicate the same distinction but are subject each to a few exceptions.

- 5. The fascicle-sheath, deciduous or persistent.
- 6. The walls of the ray-tracheids, smooth or dentate.
- 7. The connective of the pollen-sacs, large or small.
- 8. The formation of bark, late or early.

## SUBSECTIONAL CHARACTERS

An exact subdivision of the Soft Pines is possible on the following characters.

1. The umbo of the cone-scales, terminal or dorsal.
2. The scales of the conelet, mutic or armed.
3. The pits of the ray-cells, large or small.

## EVOLUTIONAL CHARACTERS

The progressive evolution of the fruit of Pinus, from a symmetrical cone of weak tissues, bearing a wingless seed, to an indurated oblique cone with an elaborate form of winged seed and an intermittent dissemination, appears among the species in various degrees of development as follows —

The seed

1. wingless.
2. with a rudimentary wing.
3. with an effective adnate wing.
4. with an ineffective articulate wing.
5. with an effective articulate wing.
6. with an articulate wing, thickened at the base of the blade.

The cone

1. indehiscent.
2. dehiscent and deciduous.
3. dehiscent and persistent.
4. persistent and serotinous.

and as to its form

1. symmetrical.
2. subsymmetrical.
3. oblique.

These different forms of the seed and, to some extent, of the cone, are available for segregating the species into groups of closely related members; while the gradual progression of the fruit, from a primitive to a highly specialized form of cone and method of dissemination, points to a veritable taxonomic evolution which is here utilized as the fundamental motive of the systematic classification of the species.

**SPECIFIC CHARACTERS**

All aspects of vegetative and reproductive organs may contribute toward a determination of species, but the importance of each character is often relative, being conclusive with one group of species, useless with another. Characters considered by earlier authors to be invariable with species, such as the dimensions of leaf or cone, the number of leaves in the fascicle, the position of the resin-ducts, the presence of pruinose branchlets, etc., prove to be inconstant in some species. In fact, as the botanical horizon enlarges, the varietal limits of the species broaden and many restrictions imposed by earlier systems are gradually disappearing.

21

Variation is the preliminary step toward the creation of species, which come into being with the elimination of intermediate forms. Variation in a species may be the result of its participation in the evolutionary processes culminating in the serotinous Pines, or it may result from the ability of the species to adapt itself to various environments by sympathetic modifications of growth, or it may arise from some peculiarity of the individual tree.

Evolutionary variation is associated with the gradual appearance of the persistent, the oblique and the serotinous cone, and of the multinodal spring-shoot. For these conditions appear in less or greater prevalence among the species of the genus.

Variation induced by environment finds familiar illustrations among the species that can survive at the limits of vegetation and can meet these inhospitable conditions by a radical change of all growing parts. Such variations are mainly of dimensions, but, with some species, the number of fascicle-leaves is affected and the shorter growing-season may modify the cone-tissues. In Mexico and Central America are found extremes of climate within small areas and easily within the range of dissemination from a single tree. The cause of the bewildering host of varietal forms, connecting widely contrasted extremes, seems to lie in the facile adaptability of those Pines, which are able to spread from the tropical base of a mountain to a less or greater distance toward its snow-capped summit.

The peculiarities of individual trees that induce abnormally short or long growths, the dwarf or other monstrous forms, the variegations in leaf-coloring, etc., etc., are not available for classification, for they may appear in any species, in fact in any genus of Conifers. These variations are artificially multiplied for commercial and decorative purposes. But inasmuch as they are repeated in all species and genera of the Coniferae that have been long under the observation of skillful gardeners, their significance has a broader scope than that imposed by the study of a single genus.

# PART II

# CLASSIFICATION OF THE SPECIES

The following classification is based on the gradual evolution of the fruit from a cone symmetrical in form, parenchymatous in tissue, indehiscent and deciduous at maturity, releasing its wingless seed by disintegration—to a cone oblique in form, very strong and durable in tissue, persistent on the tree, intermittently dehiscent, releasing its winged seeds partly at maturity, partly at indefinite intervals during several years. This evolution embraces two extreme forms of fruit, one the most primitive, the other the most elaborate, among Conifers.

Two sections of the genus, Soft and Hard Pines, are distinguished by several correlated characters, and moreover are distinct by obvious differences in the tissues of their cones as well as in the quality and appearance of their wood.

With the Soft Pines the species group naturally under two subsections on the position of the umbo, the anatomy of the wood and the armature of the conelet. In one subsection (Cembra) are found three species, P. cembra and its allies, with the cone-tissues so completely parenchymatous that the cones cannot release the seeds except by disintegration. In both subsections there is a gradual evolution from a wingless nut to one with an effective wing, adnate in one subsection, adnate and articulate in the other. The different stages of this evolution are so distinct that the Soft Pines are easily separated into definite groups.

Among the Hard Pines a few species show characters that are peculiar to the Soft Pines. These exceptional species form a subsection (Parapinaster) by themselves.

With the remaining species, the majority of the Pines, the distinctions that obtain among Soft Pines have disappeared. The dorsal umbo, the articulate seed-wing, the persistent fascicle-sheath, the dorsal and ventral stomata of the leaf and its serrate margins, the dentate walls of the ray-tracheids have become fixed and constant.

But a new form of seed-wing appears, with a thickened blade, assuming such proportions in P. Sabiniana and its two allies that these three constitute a distinct group, remarkable also for the size of its cones.

Here also appear a new form of fruit, the oblique cone, and a new method of dissemination, the serotinous cone. Associated with the latter are the persistent cone and the multinodal spring-shoot. These characters do not develop in such perfect sequence and regularity that they can be employed for grouping the species without forcing some of them into unnatural association. The oblique cone first appears sporadically here and there and without obvious reason. The persistent cone, the first stage of the serotinous cone, is equally sporadic in the earlier stages of evolution. The same may be said of the multinodal shoot.

Nevertheless these characters show an obvious progression toward a definite goal, where they are all united in a small group of species remarkable for the form and texture of their cones, for a peculiar seed-release and for the vigor and rapidity of their growth. It is possible, with the assistance of other characters, to segregate these species in three groups in which the affinities are respected and the general trend of their evolution is preserved.

The first group, the Lariciones, contains species with large ray-pits, cones dehiscent at maturity, and uninodal spring-shoots. They are, with two exceptions, P. resinosa and P. tropicalis, Old World species.

23

The second group, the Australes, contains species with small ray-pits, cones dehiscent at maturity and spring-shoots gradually changing, among the species, from a uninodal to a multinodal form. They are, without exception, species of the New World.

The third group, the Insignes, contains the serotinous species. The ray-pits are small and the spring-shoots are, with two exceptions, multinodal. With two exceptions, P. halepensis and P. pinaster, they are New World species.

These three groups, being the progressive sequence of a lineal evolution, are not absolutely circumscribed, but are more or less connected through a few intermediate species of each group. The systematic position of these intermediate species is determined by their obvious affinities. It cannot be expected that the variations, which take an important part in the evolution of the species, progress with equal step or in perfect correlation with each other.

As to specific determinations, a little experience in the field discloses an amount of variation in species that does not always appear in the descriptions of authors; and species that are under the closest scrutiny of botanists, foresters or horticulturalists, attest by their multiple synonymy their wide variation. The possibilities of variation are indefinite and, with adaptable Pines, the range of variation is somewhat proportionate to change of climate. In mountainous countries, where there are warm sheltered valleys with rich soil below cold barren ledges, the most variable Pines are found. The western species of North America, for instance, are much more variable than the eastern species, while in Mexico, a tropical country with snow-capped mountains, the variation is greatest.

Therefore in the limitation of species undue importance should not be given to characters responsive to environment, such as the dimensions of leaf or cone, the number of leaves in the fascicle, etc. Moreover, there are familiar examples (P. sylvestris, etc.) that show the possibility of wide differences in the cone of the same species.

In the following classification species only are considered without attempting to determine varietal or other subspecific forms. But varieties are often mentioned as one of the factors illustrating the scope of species. Synonymy serves a like purpose, but synonyms not conveying useful information are omitted, Roezl's list of Mexican species, for instance, and variations in the orthography of specific names.

## PINUS

- 1755 Pinus Duhamel, Traité des Arbres, ii. 121.
- 1790 Apinus Necker, Elem. Bot. iii. 269.
- 1852 Cembra Opiz, Seznam, 27.

- 1854 Strobus Opiz, Lotos, iv. 94.
- 1903 Caryopitys Small, Fl. Southeast. U. S. 29.

Leaves and shoots dimorphous, primary leaves on long shoots, secondary leaves on dwarf shoots. Flowers diclinous, the pistillate taking the place of long shoots, the staminate taking the place of dwarf shoots. Growth of wood and fruit emanating from the nodes; buds, branchlets and cones, therefore, in verticillate association. Leaves and staminate flowers in internodal position, the primary leaves along the whole length of the internode, subtending secondary leaf-fascicles on the apical, staminate flowers on the basal part. Buds compounded of minute buds in the axils of bud-scales, becoming the bracts of the spring-shoot. Branchlets of one or more internodes, each internode in three parts—a length without leaves, a length bearing leaves and a node of buds. Cone requiring two, rarely three years to mature, displaying its annual growths by distinct areas on each scale. Seeds wingless or winged, edible and nutritious.

The Pines are confined to the northern hemisphere, but grow in all climates and under all conditions of soil, temperature and humidity where trees can grow. Some of the species are of very restricted range, but others are adaptable and can cover wide areas. The sixty-six species are distributed as follows—

Eastern Hemisphere, 23.

- 1 exclusively African (Canary Islands).
- 2 exclusively European.
- 3 about the Mediterranean Basin.
- 2 common to Europe and northern Asia.
- 14 exclusively Asiatic.

Western Hemisphere, 43.

- 28 in western North America, of which 12 are confined to Mexico and Central America.

- 15 in eastern North America, of which 2 are exclusively West Indian.

The two sections of the genus correspond with those of Koehne (Deutsch. Dendrol. 28 [1893]) and his two names, Haploxylon and Diploxylon, are adopted here, together with his two subsections of Haploxylon, Cembra and Paracembra.

Of the two subsections of Diploxylon, Pinaster has been employed by Endlicher (Syn. Conif. 166 [1847]) and later authors for smaller or larger groups of Hard Pines. The subsection Parapinaster is now proposed.

The names of groups, Cembrae, Strobi, Cembroides, Gerardianae, Balfourianae, Pineae, Lariciones and Australes, are taken from Engelmann's Revision of the Genus Pinus (Trans. Acad. Sci. St. Louis, iv. 175-178 [1880]). The remainder, Flexiles, Leiophyllae, Longifoliae, Insignes and Macrocarpae, are here proposed.

In order to bring the illustrations within the limits of the page the dimensions of cone and leaf, as shown on the plates, are a little smaller than life. In plates X and XXV the reproductions of the cones are reduced to 2/5 life-size.

25

## SECTIONS, SUBSECTIONS, AND GROUPS

| | |
|---|---|
| Bases of the fascicle-bracts non-decurrent | A— HAPLOXYLON |
| Umbo of the cone-scale terminal | a— Cembra |
| Seeds wingless. | |
| Cones indehiscent | I. Cembrae |
| Cones dehiscent | II. Flexiles |
| Seed with an adnate wing | III. Strobi |
| Umbo of the cone-scale dorsal | b— Paracembra |
| Seeds wingless | IV. Cembroides |

| | |
|---|---|
| Seed-wing short, ineffective | V. Gerardianae |
| Seed-wing long, effective | VI. Balfourianae |
| | |
| Bases of the fascicle-bracts decurrent | B—DIPLOXYLON |
| | |
| Fascicle-sheath or seed of Haploxylon | c— Parapinaster |
| Fascicle-sheath deciduous | VII. Leiophyllae |
| Fascicle-sheath persistent. | |
| Seed-wing of the Strobi | VIII. Longifoliae |
| Seed-wing of the Gerardianae | IX. Pineae |
| | |
| Fascicle-sheath persistent, seed-wing articulate, effective | d— Pinaster |
| Base of wing-blade thin or slightly thickened. | |
| Cones dehiscent at maturity. | |
| Pits of ray-cells large | X. Lariciones |
| Pits of ray-cells small | XI. Australes |
| Cones serotinous, pits of ray-cells small | XII. Insignes |
| Base of wing-blade very thick | XIII. Macrocarpae |

26

# HAPLOXYLON

Bases of the bracts subtending leaf-fascicles not decurrent. Staminate flowers not sufficiently developed in the bud to be apparent. Spring-shoots uninodal. Fibro-vascular bundle of the leaf single. Cone symmetrical, of relatively fewer larger scales, its tissues softer. Bark-formation late, the trunks of young trees smooth. Wood soft and with little resin, of uniform color and with relatively obscure definition of the annual rings. Tracheids of the medullary rays with smooth walls.

All the species of this section, except P. Nelsonii, have deciduous fascicle-sheaths. There are but two species of Diploxylon with deciduous sheaths, P. leiophylla and P. Lumholtzii, both of them easily recognized. The deciduous sheath, therefore, is an obvious and useful means for recognizing the Soft Pines. On the characters of the fruit and the wood Haploxylon can be divided into two subsections.

a. Cembra             Umbo of the cone-scale terminal.
b. Paracembra         Umbo of the cone-scale dorsal.

## Cembra

Umbo of the cone-scale terminal. Scales of the conelet unarmed. Leaves in fascicles of 5, the sheath deciduous, the two dermal tissues distinct, the hypoderm-cells uniform. Pits of the cells of the wood-rays large.

Seeds wingless.
Cones indehiscent                    I. Cembrae.
Cones dehiscent                      II. Flexiles.
Seeds with an adnate wing            III. Strobi.

## I. CEMBRAE

Seeds wingless. Cones indehiscent, deciduous at maturity.

In this group of species there is no segregation of sclerenchyma into an effective tissue. The cones are inert under hygrometric changes and may always be recognized in herbaria by their persistent occlusion and soft tissues. The seeds are released only by the disintegration of the fallen cone. There is, however, a vicarious dissemination by predatory crows (genus Nucifraga) and rodents.

Leaves serrulate, their stomata ventral only.
Cones relatively larger, the apophyses protuberant   1. koraiensis.
Cones relatively smaller, the apophyses appressed    2. cembra.
Leaves entire, their stomata ventral and dorsal      3. albicaulis.

## 1. PINUS KORAIENSIS

- 1784 P. strobus Thunberg, Fl. Jap. 275 (not Linnaeus).
- 1842 P. koraiensis Siebold & Zuccarini, Fl. Jap. ii. 38.

- 1857 P. mandschurica Ruprecht in Bull. Acad. Sci. St. Pétersb. xv. 382.

Spring-shoots more or less densely tomentose. Leaves from 8 to 12 cm. long, serrulate, stomata ventral only, resin-ducts medial and confined to the angles. Conelets large, subterminal, or on young trees often pseudolateral. Cones indehiscent, from 9 to 14 cm. long, short-pedunculate, ovoid-conical or subcylindrical; apophyses dull pale nut-brown, rugose, shrinking much in drying and 27 exposing the seeds, prolonged and tapering to a more or less reflexed tip, the umbo inconspicuous; seeds large, wingless, the spermoderm entire.

A species of the mountains of northeastern Asia with valuable wood and large edible nuts; hardy and often cultivated in cool-temperate climates.

The P. koraiensis of Beissner (in Nuov. Giorn. Bot. Ital. n. ser. iv. 184) and of Masters (in Gard. Chron. ser. 3, xxxiii. 34, ff.) are P. Armandi and have led to an erroneous extension of the range of this species into Shensi and Hupeh. In the original description of the species the authors call attention to an error in the plate, where a cone of another species has been substituted.

P. koraiensis resembles P. cembra in leaf and branchlet but not in the cone. It is often confused with P. Armandi, but can easily be distinguished by its tomentose branchlets, indehiscent cone and peculiar seed. The two species, moreover, do not always agree in the position of the foliar resin-ducts.

Plate VIII.

Fig. 85, Cone and seed. Fig. 86, Leaf-fascicle and magnified leaf-section.

## 2. PINUS CEMBRA

- 1753 P. cembra Linnaeus, Sp. Pl. 1000.
- 1778 P. montana Lamarck, Fl. Franç. iii. 651 (not Miller).
- 1858 P. pumila Regel in Index Sem. Hort. Petrop. 23.
- 1884 P. mandschurica Lawson, Pinet. Brit. i. 61, ff. (not Ruprecht).

- 1906 P. sibirica Mayr, Fremdl. Wald- & Parkb. 388.
- 1913 P. coronans Litvinof in Trav. Mus. Bot. Acad. St. Pétersb. xi. 23, f.

Spring-shoots densely tomentose. Leaves from 5 to 12 cm. long, serrulate; stomata ventral only; resin-ducts medial or, in the dwarf form, often external. Conelets short-pedunculate, purple during their second season. Cone from 5 to 8 cm. long, ovate or subglobose, subsessile; apophyses dull nut-brown, thick, slightly convex, the margin often a little reflexed, the umbo inconspicuous; seeds wingless, large, the dorsal spermoderm adnate partly to the nut, partly to the cone-scale, the ventral spermoderm wanting.

The Swiss Stone Pine attains a height of 15 or 25 metres and occupies two distinct areas, the Alps, from Savoy to the Carpathians at high altitudes, and the plains and mountain-slopes throughout the vast area from northeastern Russia through Siberia. Beyond the Lena and Lake Baikal it becomes a dwarf (var. pumila) with its eastern limit in northern Nippon and in Kamchatka. It is successfully cultivated in the cool-temperate climates of Europe and America. The wood is of even, close grain, peculiarly adapted to carving. The nuts are gathered for food and confections, but are destroyed in great numbers by squirrels, mice and a jay-like crow, the European Nutcracker. It is generally conceded, however, that these enemies assist in dissemination.

Plate VIII.

Fig. 87, Cone, seed and magnified leaf-section. Fig. 88, Tree at Arolla, Switzerland. Fig. 89, Cone, leaf-fascicle and magnified leaf-section of var. pumila.

## 3. PINUS ALBICAULIS

- 1853 P. flexilis Balfour in Bot. Exped. Oregon, 1, f. (not James).
- 1857 P. cembroides Newberry in Pacif. R. R. Rep. vi-3, 44, f. (not Zuccarini).
- 1863 P. albicaulis Engelmann in Trans. Acad. Sci. St. Louis, ii. 209.

- 1867 P. shasta Carrière, Trait. Conif. ed. 2, 390.

Spring-shoots glabrous or pubescent. Branchlets pliant and tough. Leaves from 4 to 7 cm. long, entire, stout, persistent for several years; stomata dorsal and ventral; resin-ducts external. Conelets short-pedunculate, dark purple during the second season, their scales often tapering to an acute apex. Cones from 5 to 7 cm. long, subsessile, oval or subglobose; apophyses nut-brown or fulvous brown, dull or slightly lustrous, very thick, the under surface conspicuous, meeting the upper surface 28 in an acute margin, and terminated by a salient, often acute umbo; seed wingless, the testa bare of spermoderm.

This species ranges from British Columbia through Washington and Oregon, over the mountains of northern California and the Sierras as far south as Mt. Whitney, and, on the Rocky Mountains, through Idaho and Montana to northern Wyoming. It is found at the timber-line of many stations and forms, in exposed situations, flat table-like masses close to the ground. It is a species of no economical importance and is too inaccessible for the profitable gathering of its large nuts, which are devoured in quantity by squirrels and by Clark's crow, a bird of the same genus with the pinivorous Nutcracker of Europe.

P. albicaulis is distinguished from its allies by its entire leaves with both dorsal and ventral stomata, from P. flexilis by its indehiscent cone, and from all of these species by its seed without membranous cover or rudimentary wing. It was united with P. flexilis by Parlatore and Gordon, and, later, was referred to that species as a varietal form by Engelmann (in Brewer & Watson, Bot. Calif. ii. 124). Parrish's P. albicaulis (in Zoe, iv. 350), extending its range to the mountains of southern California, proves to be P. flexilis (Jepson, Silva Calif. 74).

Plate VIII.

Fig. 90, Two cones and seed. Fig. 91, Leaf-fascicle. Fig. 92, Magnified leaf-section.

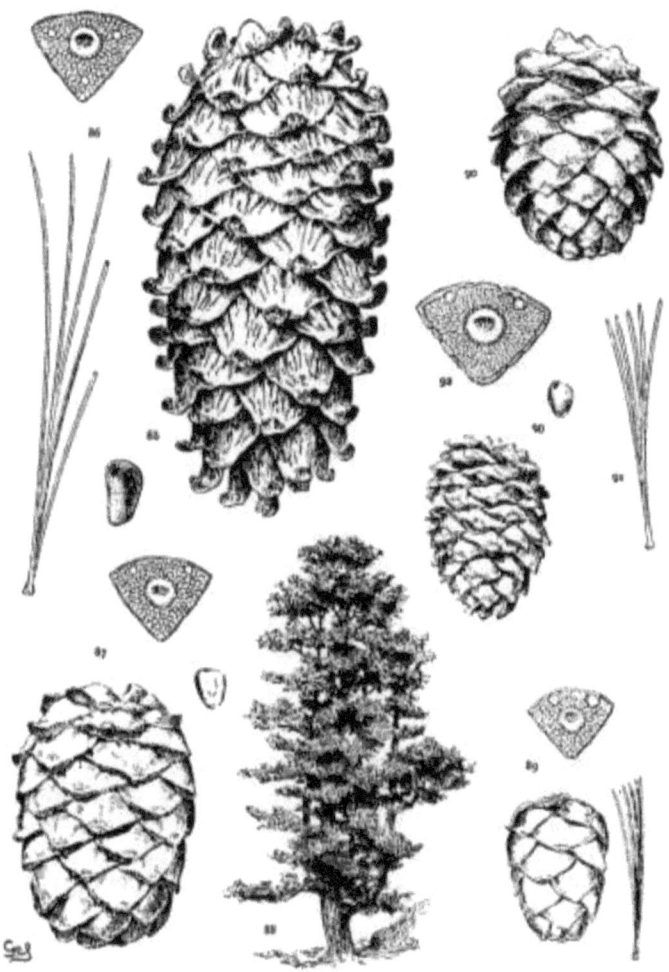

PLATE VIII. P. KORAIENSIS (85, 86), CEMBRA (87-89), ALBICAULIS (90-92)

## II. FLEXILES

Seeds wingless, the spermoderm forming a narrow border with a rudimentary prolongation. Cones dehiscent at maturity.

The dehiscent cone distinguishes this group from the Cembrae. Therefore confusion of P. koraiensis with P. Armandi, or P. albicaulis with P. flexilis should be impossible. The peculiar seed is found again only in the northern variety of P. ayacahuite.

Leaves usually entire, the stomata dorsal and ventral   4. flexilis.

Leaves serrulate, the stomata ventral only   5. Armandi.

## 4. PINUS FLEXILIS

- 1823 P. flexilis James in Long's Exped. ii. 34.
- 1882 P. reflexa Engelmann in Bot. Gaz. vii. 4.
- 1897 P. strobiformis Sargent, Silva N. Am. xi. 33, tt. 544, 545 (not Engelmann).

Spring-shoots pubescent; branchlets very tough and pliant. Leaves from 3 to 9 cm. long, entire, or serrulate in the southern variety, persistent for five or six years; stomata dorsal and ventral or, in the south, sometimes ventral only; resin-ducts external. Cones from 6 to 25 cm. long, ovate or subcylindrical, short-pedunculate; apophyses pale tawny yellow, or yellow ochre, lustrous, often prolonged and more or less reflexed, thick, the margin together with the umbo raised above the surface of the cone.

This species grows on the Rocky Mountains from Alberta in the Dominion of Canada to Chihuahua in northern Mexico and ranges westward to the eastern slope of the Sierras and to the southern mountains of California. The wood, where accessible, is manufactured into lumber. It may be seen in the Arnold Arboretum and in the Royal Gardens at Kew.

P. flexilis is recognized by its lustrous yellow cones. This and the constantly external ducts of its usually entire leaves distinguish it from P. Armandi. From P. albicaulis, with similar leaves, it differs by its dehiscent cone. At one extreme the cone of P. flexilis is not unlike that of P. albicaulis, at the other extreme it approaches the characteristic cone of P. ayacahuite, with prolonged reflexed scales. Hence the confusion of P. albicaulis with P. flexilis (Murray, Parlatore and others) and of P. flexilis with Engelmann's P. strobiformis. Sargent's P. strobiformis, illustrated in the Silva of North America, is the form of this species known as var. reflexa of Engelmann.

Plate IX.

Fig. 93, Two cones and seed. Fig. 94, Leaf-fascicle. Fig. 95, Magnified leaf-section.

30

## 5. PINUS ARMANDI

- 1884 P. Armandi Franchet in Nouv. Arch. Mus. Paris, sér. 2, vii. 95, 96, t. 12.
- 1898 P. scipioniformis Masters in Bull. Herb. Boiss. vi. 270.
- 1903 P. koraiensis Masters in Gard. Chron. ser. 3, xxxiii. 34, ff. 18, 19 (not Siebold & Zuccarini).
- 1908 P. Mastersiana Hayata in Gard. Chron. ser. 3, xliii, 194.

Spring-shoots glabrous; branches and most of the trunk covered with a smooth gray cortex. Leaves from 8 to 15 cm. long, serrulate; stomata ventral only; resin-ducts external, external and medial, or medial, all three conditions sometimes occurring in leaves of the same branchlet. Cones from 6 to 20 cm. in length, pendent on peduncles of various lengths, the peduncle often remaining on the tree after the fall of the cone; apophyses fulvous brown, dull or sublustrous, the margin rounded or tapering to an acute apex, sometimes a little prolonged and reflexed, the umbo inconspicuous.

A tree of the mountains of central, southern and western China with an outlying station on the Island of Formosa. Recently planted in Europe and America, it has so far proved hardy. The nuts are gathered for food and some use is made of the wood.

The glabrous shoots of P. Armandi distinguish it from P. flexilis and P. koraiensis. From the latter it is also distinct in its dehiscent cone and in its seed. The section of its leaf, with dorsal ducts often in two positions, is peculiar to this species among Soft Pines.

Plate IX.

Fig. 96, Two cones and seed. Fig. 97, Leaf-fascicle. Figs. 98, 99, Magnified sections of three leaves.

31

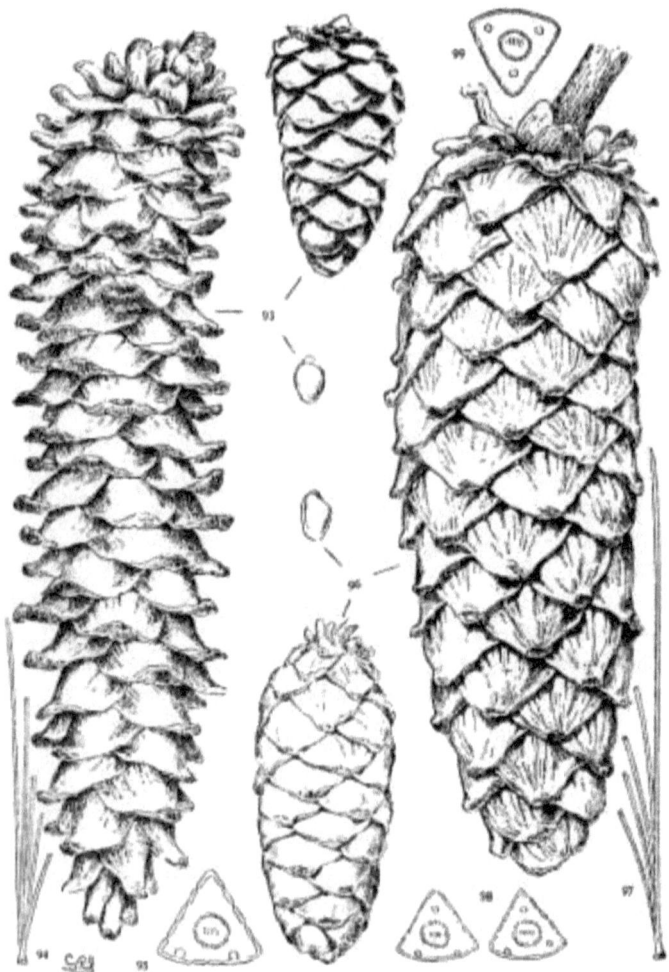

PLATE IX. P. FLEXILIS (93-95), ARMANDI (96-99)

## III. STROBI

Seed with a long effective wing adnate to the nut.

The base of the seed-wing corresponds to the marginal spermoderm of the Flexiles but is prolonged into an effective adnate wing. This form of wing appears again in the species Balfouriana and in the group Longifoliae.

Cones very long, usually exceeding 25 cm.
| | |
|---|---|
| Cone-scales prolonged and reflexed | 6. ayacahuite. |
| Cone-scales appressed | 7. Lambertiana. |

Cones less than 25 cm. long.
Cone-scales prominently convex.
| | |
|---|---|
| Leaves less than 7 cm. long | 8. parviflora. |
| Leaves 9-12 cm. long | 9. peuce. |
| Leaves 12-18 cm. long | 10. excelsa. |

Cone-scales thin, conforming to the surface of the cone.
| | |
|---|---|
| Cone relatively longer, its phyllotaxis 8/21 | 11. monticola. |
| Cone relatively shorter, its phyllotaxis 5/13 | 12. strobus. |

## 6. PINUS AYACAHUITE

- 1838 P. ayacahuite Ehrenberg in Linnaea, xii. 492.
- 1848 P. strobiformis Engelmann in Wislizenus, Tour Mex. 102.
- 1857 P. Veitchii Roezl, Cat. Graines Conif. Mex. 32.
- 1858 P. Bonapartea Roezl in Gard. Chron. 358.
- 1858 P. Loudoniana Gordon, Pinet. 230.

Spring-shoots glabrous or pubescent. Leaves from 10 to 20 cm. long, serrulate, their stomata ventral only, their resin-ducts external, often numerous. Cones from 25 to 45 cm. long, pendent on long stalks, subcylindrical or tapering, often curved; apophyses pale nut-brown, dull or sublustrous, varying much in thickness, prolonged in various degrees, the prolongations patulous, reflexed, 32 recurved or revolute; seeds of the southern typical form with a long wing, the wing diminishing and the nut increasing in relative size northward.

The White Pine of Mexico and Guatemala grows on mountain-slopes and at the head of ravines. It is not very hardy in cultivation except in the milder parts of Great Britain and in northern Italy, where the forms of central and northern Mexico have been very successful. The species is best recognized by the prolonged apophyses of its large cone.

The variations in the size of the cone and in the prolongations of its scales are many, but of far more significance is the remarkable variation of the seed-wing, which is long in the southern part of the range, short and broad in central Mexico, and rudimentary, like the seed of P. flexilis, in the north. This makes it possible to establish two well defined varieties—Veitchii and brachyptera. The three forms of the species present a gradation from the long effective wing of the Strobi to the rudimentary form of the Flexiles. Many of the seed-wings of the var. Veitchii correspond, in their short broad form and opaque coloring, with the characteristic wing of P. Lambertiana.

Plate X. (leaves and cones much reduced).

Fig. 103, Cone and cone-scale of var. Veitchii. Fig. 104, Cone and seed of var. brachyptera. Fig. 105, Cone-scale of the typical form. Figs. 106, 107, Leaf-fascicles and magnified leaf-sections.

## 7. PINUS LAMBERTIANA

- 1827 P. Lambertiana Douglas in Trans. Linn. Soc. xv. 497.

Spring-shoots pubescent. Leaves from 7 to 10 cm. long, serrulate; stomata dorsal and ventral; resin-ducts external or with one or two ventral medial ducts. Cones from 30 to 50 cm. long, pendent, subcylindrical, tapering to a rounded apex; apophyses pale nut-brown, thick, a narrow border of the under surface showing on the closed cone, the margin rounded or tapering to a blunt slightly reflexed tip; seed with a large nut and a broad short opaque wing.

The Sugar Pine is the tallest of the genus and attains a height of 50 or 60 metres. It grows on mountain slopes and the sides of ravines. Its southern limit is in Lower California on the plateau of San Pedro Martir, its northern limit is in western Oregon. The wood is valuable, its nuts are eaten by native Indians, and the sweet exudation, which gives the tree its popular name, is a manna-like substance of some officinal value. P. Lambertiana is recognized by its long cone and by the constant dorsal stomata of its leaves.

Plate X. (leaves and cone much reduced).

Fig. 100, Cone and seed. Fig. 101, Conelet. Fig. 102, Leaf-fascicle and magnified leaf-section.

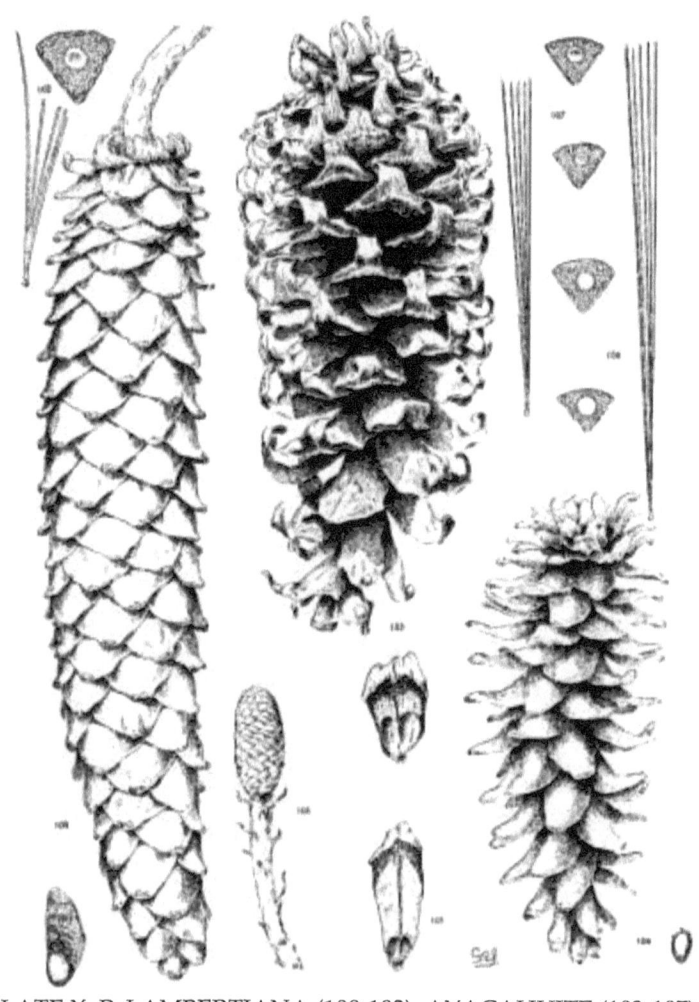

PLATE X. P. LAMBERTIANA (100-102), AYACAHUITE (103-107)

## 8. PINUS PARVIFLORA

- 1784 P. cembra Thunberg, Fl. Jap. 274. (not Linnaeus).
- 1842 P. parviflora Siebold and Zuccarini, Fl. Jap. ii. 27, t. 115.

- 1890 P. pentaphylla Mayr, Mon. Abiet. Jap. 78, 94, t. 6.
- 1908 P. morrisonicola Hayata in Gard. Chron. ser. 3, xliii. 194.
- 1908 P. formosana Hayata in Jour. Linn. Soc. xxxviii. 297, t. 22.

Spring-shoots pubescent or glabrous; branches becoming studded with prominent resin-cells of the cortex. Leaves from 3 to 8 cm. long, slender, serrulate; stomata ventral only; resin-ducts external and dorsal. Cones subsessile, often persistent, from 5 to 10 cm. long, patulous or horizontal, short-ovate, or elongate and slightly conical; apophyses nut-brown, abruptly convex near the apex, or irregularly warped, varying much in size, the umbo confluent with the thin margin of the scale and resting on the apophysis beneath; seeds with a large nut and a short broad wing, often temporarily adherent to the cone-scale and breaking apart at the fall of the nut.

A tree of the mountains of Japan and Formosa, cultivated extensively. It is recognized by its very short quinate leaves and by its nearly sessile cones. The frequent but not invariable retention of 34 the seed-wing in the cone is due to adhesion. Many seeds fall with their wings intact, others break away from the wing which, after a while, loosens and also falls.

Plate XI.

Figs. 114, 115, Three cones and seed. Fig. 116, Leaf-fascicle and magnified leaf-section.

### 9. PINUS PEUCE

- 1844 P. peuce Grisebach, Spicil. Fl. Rumel. ii. 349.
- 1865 P. excelsa Hooker in Jour. Linn. Soc. viii. 145. (not Wallich).

Spring-shoots glabrous. Leaves from 7 to 10 cm. long, erect, serrulate; stomata ventral only; resin-ducts external. Connective of pollen-sacs small and narrow. Cones deciduous, from 8 to 15 cm. long, subcylindrical, often curved, the peduncle short; apophyses tawny yellow, prominently and abruptly convex, the umbo against the scale beneath; seed-wing long.

A tree of the Balkan Mountains, very hardy and bearing abundant fruit in the gardens of both hemispheres. The cone resembles that of P. excelsa, but is prevalently much shorter and with a relatively shorter peduncle. Its leaves are also much shorter and are always erect. A curious difference is found in the connectives of the pollen-sacs, small in peuce (fig. 113), large in excelsa (fig. 110). The convexity of its apophyses distinguishes the cone from those of P. monticola and P. strobus. Beissner followed Hooker and named this species excelsa, var. peuce, in the first edition of his Handbuch (1891), but in the second edition he restored the Balkan Pine to specific standing.

Plate XI.

Fig. 111, Cone and seed. Fig. 112, Leaf-fascicle and magnified leaf-section. Fig. 113, Pollen-sacs and connective magnified.

## 10. PINUS EXCELSA

- 1824 P. excelsa Wallich ex Lambert, Gen. Pin. ii, 5, t. 3.
- 1845 P. nepalensis De Chambray, Arbr. Résin. 342.
- 1854 P. Griffithii McClelland in Griffith, Notul. Pl. Asiat. iv, 17; Icon. Pl. Asiat. t. 365.

Spring-shoots glabrous. Leaves from 10 to 18 cm. long, drooping, serrulate; stomata ventral only; resin-ducts external but often with a medial ventral duct. Connective of the pollen-sacs large. Cones from 15 to 25 cm. long, narrow-cylindrical; apophyses tawny yellow or pale fulvous brown, prominently convex, the umbo against the apophysis beneath; seeds with a long wing.

A tree with gray-green drooping foliage, found, with some interruptions, along the Himalayas. It furnishes resin, tar and wood of considerable value. It is cultivated in all temperate climates and is a familiar tree of American and European gardens. Madden states that the foliage of P. excelsa is sometimes erect and is occasionally bright green. Such variations are often met in other species of Pinus.

Usually the drooping gray-green foliage and the peculiar cone are sufficient for the recognition of this species. The not infrequent presence of a medial duct and the large connective are valuable aids for identifying it.

Plate XI.

Fig. 108, Cone and seed. Fig. 109, Leaf-fascicle and magnified section of two leaves. Fig. 110, Pollen-sacs and connective magnified.

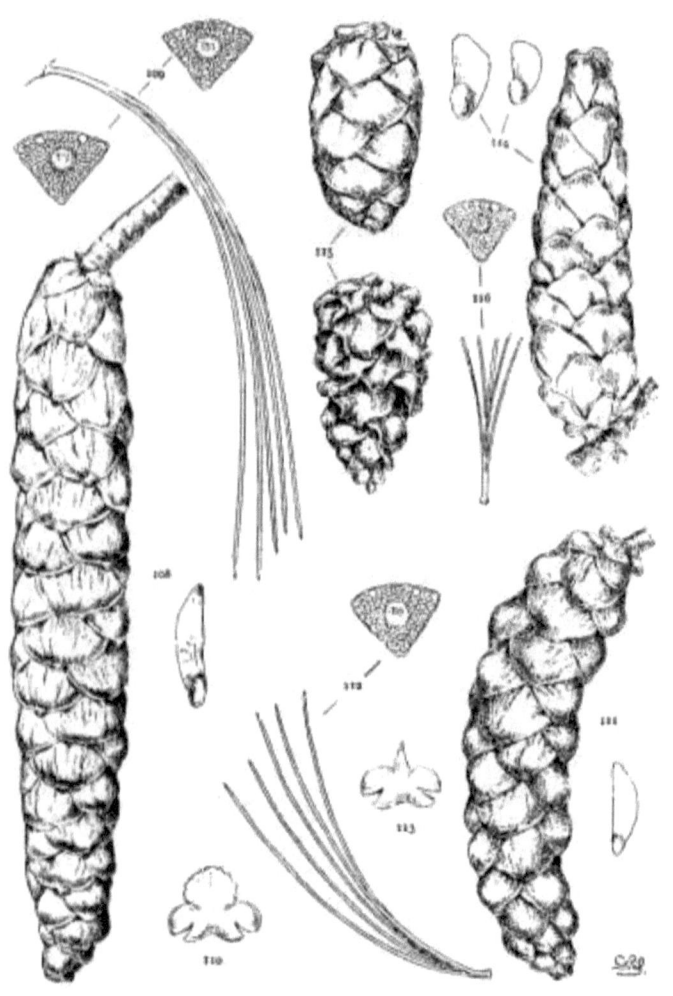

PLATE XI. P. EXCELSA (108-110), PEUCE (111-113), PARVIFLORA (114-116)

## 11. PINUS MONTICOLA

- 1837 P. monticola Douglas ex Lambert, Gen. Pin. iii. t.
- 1884 P. porphyrocarpa Lawson, Pinet. Brit. i, 83, ff.

Spring-shouts pubescent. Leaves from 4 to 10 cm. long, serrulate; stomata ventral or rarely with a few dorsal stomata; resin-ducts external. Cones from 10 to 25 cm. long, cylindrical or tapering, sometimes curved; apophyses brown-ochre or fulvous brown, thin, smooth, conforming to the surface of the cone, the apex sometimes slightly prolonged and reflexed, the umbo not quite touching the surface of the scale below.

36

The western White Pine grows in southern British Columbia and on Vancouver Island, on the Rocky Mountains of Montana and Idaho, in Washington, on the Blue Mountains, Cascades and Coast Range of Oregon, across northern California and along the Sierras to the mountains of southern California. Where it is abundant and accessible it furnishes valuable timber. It is hardy in New England and in northern and central Europe.

It differs from P. strobus in the higher phyllotaxis of its cone, an obvious difference that may be seen by comparing cones of the two species of the same length (figs. 117, 119), the number of scales on the cone of P. monticola being much greater than that on the cone of P. strobus. Nuttall (Sylva, iii, 118) followed Hooker in considering it to be a variety of P. strobus.

Plate XII.

Fig. 117, Cone and cone-scale. Fig. 118, Leaf-fascicle and magnified leaf-section.

## 12. PINUS STROBUS

- 1753 P. strobus Linnaeus, Sp. Pl. 1001.
- 1855 P. nivea Booth ex Carrière, Trait. Conif. 305.
- 1862 P. alba-canadensis Provancher, Fl. Canad. ii. 554.
- 1903 Strobus strobus Small, Fl. Southeast. U. S. 29.

Spring-shoots pubescent. Leaves from 6 to 14 cm. long, serrulate; stomata ventral only; resin-ducts external. Cones from 8 to 24 cm. long, narrow cylindrical, sometimes curved; apophyses fulvous brown, or rufous brown, thin, the smooth or slightly rugose surface conforming to the general surface of the cone; seed with a long wing.

A valuable timber-tree of singular beauty and rapid growth. The northern limit of its range extends from Newfoundland to Manitoba; it grows throughout the northern states from Minnesota to the Atlantic, and, south of Pennsylvania, along the Appalachians to northern Georgia. Its tractable and reliable wood, its adaptability to various soils and climates, its early maturity and stately habit, recommend it to the forester and gardener.

Mature trees of P. strobus tower above the evergreens associated with it. It is also recognized by the color and horizontal massing of its foliage. The cone, when closed, is very narrow; its thin flat scales distinguish it from the cone of P. peuce, and its phyllotaxis from the cone of P. monticola. To illustrate the possibilities of variation in the size of Pine cones, I once collected several in Tamworth, N. H., on the estate of Mr. Augustus Hemenway, on the same slope and within an area of one square kilometre. These cones varied in length from 6 to 24 cm., with all intermediate sizes. Also on each tree were cones of various lengths, but the longest were confined to two or three trees among the several hundred examined. Dimensions of leaves also varied with individual trees; not infrequently the leaves of a tree were twice the length of those of an adjacent tree. Such variations appear in many species and in many localities.

Plate XII.

Fig. 119, Two cones. Fig. 120, Leaf-fascicle. Fig. 121, Magnified leaf-section. Fig. 122, Conelets. Fig. 123, A cultivated tree in Massachusetts.

37

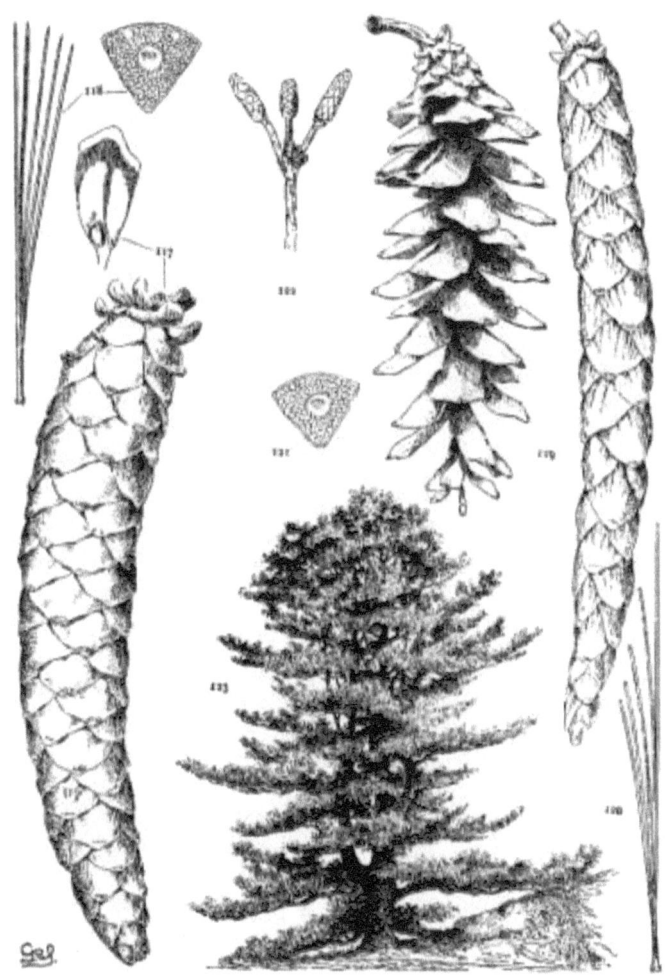

PLATE XII. P. MONTICOLA (117, 118), STROBUS (119-123)

### Paracembra

Umbo of the cone-scale dorsal. Scales of the conelet mucronate or aristate. Epiderm and hypoderm of the leaf similar, appearing as a single tissue; resin-ducts external. Pits of the ray-cells small.

The wood of this subsection differs from that of other species, except that of P. pinea, in the Picea-like characters of the medullary rays—tracheids with smooth walls combined with the thick walls and small pits of the ray-cells. On the character of the seeds the species may be divided into three groups.

| | |
|---|---|
| Seeds wingless | IV. Cembroides. |
| Seeds with a short, ineffective, articulate wing | V. Gerardianae. |
| Seeds with a long and effective wing | VI. Balfourianae. |

38

## IV. CEMBROIDES

Seeds wingless, the nut large, wholly or partly bare of membranous cover. Cones varying from yellow-ochre to deep red-orange in color.

These are the Nut Pines, growing on the arid slopes and tablelands above the great plateau of northern Mexico and its extension into the southwestern United States. There are three distinct species.

| | |
|---|---|
| Leaves entire, the sheath deciduous. | |
| Cones subglobose, subsessile | 13. cembroides. |
| Cones cylindrical, pedunculate | 14. Pinceana. |
| Leaves serrulate, the sheath persistent | 15. Nelsonii. |

## 13. PINUS CEMBROIDES

- 1832 P. cembroides Zuccarini in Abh. Akad. Münch. i. 392.
- 1838 P. Llaveana Schiede in Linnaea, xii. 488.
- 1845 P. monophylla Torrey in Frémont's Rep. 319, t. 4.
- 1847 P. Fremontiana Endlicher, Syn. Conif. 183.
- 1848 P. edulis Engelmann in Wislizenus, Tour. Mex. 88.
- 1848 P. osteosperma Engelmann in Wislizenus, Tour. Mex. 89.
- 1862 P. Parryana Engelmann in Am. Jour. Sci. ser. 2, xxxiv. 332 (not Gordon).
- 1897 P. quadrifolia Sudworth, Bull. 14, U. S. Dep. Agric. 17.

- 1903 Caryopitys edulus Small, Fl. Southeast. U. S. 29.

Spring-shoots pruinose. Leaves from 2 to 6 cm. long, in fascicles of 1 to 5, the sheath-scales revolute at the apex, then deciduous; stomata ventral, or ventral and dorsal; resin-ducts external. Scales of the conelet armed with a minute prickle. Cones from 4 to 6 cm. long, subglobose, subsessile; apophyses lustrous ochre-yellow, crowned with a quadrilateral umbo bearing the minute prickle of the conelet; seed flaxen yellow when fresh, its testa bare, the spermoderm adnate to the cone-scale.

A broad tree with a round head, similar in size and form, but not in ramification, to the cultivated Apple-tree; growing on arid slopes and table-lands. Its eastern limit is in southwestern Wyoming, central Colorado, Texas, western Tamaulipas and northwestern Vera Cruz. It ranges over Utah, Nevada, Arizona and the northern states of Mexico to the southern Sierras of California and to the northern and southern extremities of Lower California. It is recognized by its small cone, which expands, when open, into an irregular flat aggregate of loosely attached scales. The leaves are shorter than those of the other Pines of this group.

The cone of this species always retains its peculiar character. The variations are mainly in the number of leaves in the fascicle. On this character this Nut Pine is divided by many authors into four species—cembroides, with three slender leaves—edulis, with two stout leaves—monophylla, with one leaf and—Parryana, with four stout leaves. But there are intermediate forms that may be either cembroides or edulis, edulis or monophylla etc., and Voss's reduction of the four to a single species with three varieties seems to be justified (Mitt. Deutsch. Dendrol. Ges. xvi. 95).

Plate XIII.

Fig. 130, Cone, cone-scale and seed. Fig. 131, Open cone. Fig. 132, Branchlet with leaves and magnified leaf-section.

## 14. PINUS PINCEANA

- 1846 P. cembroides Gordon in Jour. Hort. Soc. Lond. i. 236, f. (not Zuccarini).

- 1858 P. Pinceana Gordon, Pinet. 204.
- 1882 P. latisquama Engelmann in Gard. Chron. ser. 2, xviii. 712. f. 125 (as to cone only).

Spring-shoots slender, pruinose. Leaves in fascicles of three, the sheath revolute at the base, then deciduous; stomata ventral, or ventral and dorsal; resin-ducts external. Scales of the conelet minutely mucronate. Cones from 6 to 9 cm. long, cylindrical, pendent on long peduncles; apophyses 40 lustrous ochre-yellow, elevated in the centre, the umbo usually retaining the small prickle; seed large, bearing on its dorsal surface remnants of the spermoderm.

A small bushy tree with long slender branchlets, clear gray cortex, persistently smooth except on the lower part of the trunk, and glaucous-green foliage. It grows along water-courses, dry in autumn and winter, from southern Coahuila to central Hidalgo, and is associated with P. cembroides, from which it may be distinguished by its longer leaves and much longer cylindrical cone.

Plate XIII.

Fig. 127, Cone, cone-scale and seed. Fig. 128, Branchlet with leaves. Fig. 129, Magnified leaf-section.

## 15. PINUS NELSONII

- 1904 P. Nelsonii Shaw in Gard. Chron. ser. 3, xxxvi. 122, f. 49.

Spring-shoots slender, pruinose; branchlets very pliant and tough, summer-shoots abundant. Leaves with a persistent sheath, from 6 to 9 cm. long, united in threes along a portion of their ventral surface into pseudomonophyllous fascicles, serrulate on the two margins of the dorsal surface, entire on the ventral margin; stomata dorsal and with one row along the free portion of each ventral face. Conelets usually, if not always, pseudolateral by reason of the summer growth of the branchlets, and attaining in their first season an unusually large size. Cones from 6 to 12 cm. long, on very long stout and curved peduncles, cylindrical, deciduous by an articulation between the cone and its peduncle, leaving the latter for several years on the tree; apophyses dark lustrous orange-red, rugose, elevated along a sharp transverse keel, the umbo obscurely defined, the mucro usually broken away; nuts large, flaxen yellow, the spermoderm adnate to the cone-scale.

A small bushy tree with long pliant branches, clear gray cortex all over the limbs and trunk, and sparse gray-green foliage. It grows, together with P. cembroides, on the lower slopes of the northeastern Sierras of Mexico, near the boundary between the states of Tamaulipas and Nuevo Leon. It is apparently confined to a small area near the latitude of the city of Victoria, the capital of Tamaulipas, where its nuts are often exposed for sale.

In many characters this species is unique. It can be recognized at once by the connate leaves that form the fascicle or by the remarkable stout curved peduncle of its cone. Such seeds as I have seen differ from those of P. cembroides by a reddish area at one end, but this can be seen with fresh seeds only.

Plate XIII.

Fig. 124, Cone, cone scale and seed. Fig. 125, Branchlet with leaves. Fig. 126, Magnified section of a leaf-fascicle.

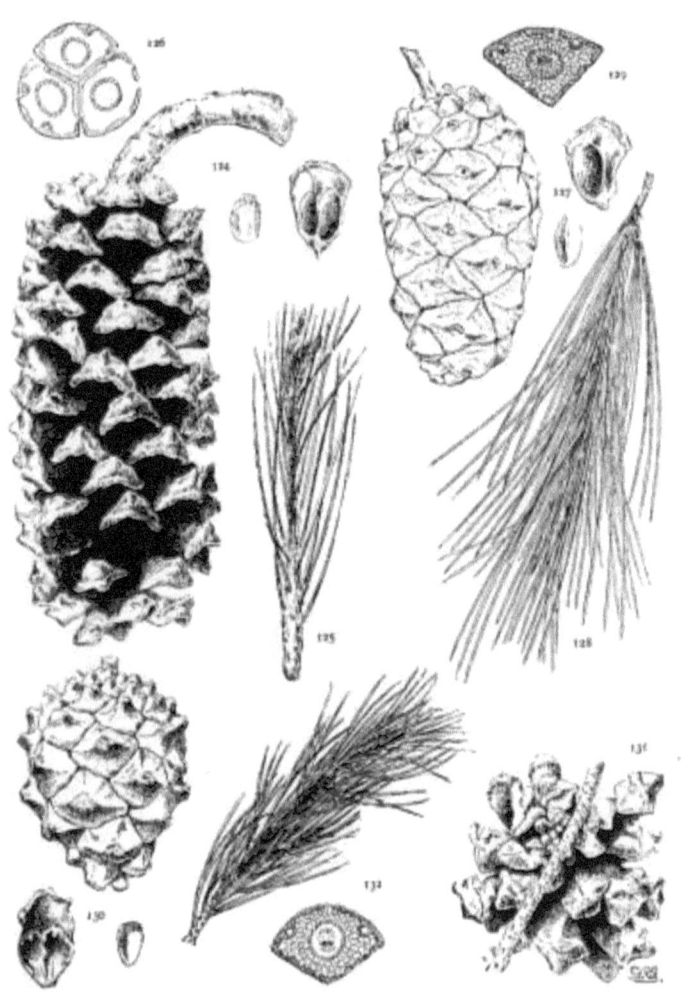

PLATE XIII. P. NELSONII (124-126), PINCEANA (127-129), CEMBROIDES (130-132)

## V. GERARDIANAE

Seeds with a very short ineffective articulate wing. Leaves in fascicles of 3, serrulate, the sheath deciduous. Bark exfoliating in large scales, leaving parti-colored areas.

These Asiatic Nut Pines are alike in leaf and cortex as well as in the peculiar seed-wing. The last often remains in the cone after the nut falls. The mechanical nature of this adhesion is apparent in P. Gerardiana, where the wing adheres not to its own, but to the adjacent scale. The two species are alike in their leaves but distinct in their cones and seeds.

| | |
|---|---|
| Cones smaller, the nut short-ovate | 16. Bungeana. |
| Cones larger, the nut long-cylindrical | 17. Gerardiana. |

## 16. PINUS BUNGEANA

- 1847 P. Bungeana Zuccarini ex Endlicher, Syn. Conif. 166.

Spring-shoots glabrous, summer-shoots common on fruiting branches of young trees. Leaves from 6 to 10 cm. long, serrulate; stomata dorsal and ventral; resin-ducts external. Conelets subterminal or often pseudolateral, their scales gradually narrowed into a spine. Cones from 5 to 7 cm. long, 42 short-pedunculate, short-ovate; apophyses dull pale nut-brown, elevated along a transverse keel, the dark brown umbo forming a spine with a broad base; seeds with a short loosely attached wing, sometimes remaining in the cone when the short-ovate nut falls.

A tree cultivated about the temples of China and recently found by Wilson growing on the mountains of Hupeh. The earlier particolored bark changes to chalky white on old trunks, by which the tree is recognized from a great distance. The stem of the tree is often multiple by the vertical growth of some of the lower branches. It is very hardy and is cultivated in Europe and America, although these cultivated trees are not yet of sufficient age to show the remarkable white trunk.

Plate XIV.

Fig. 138, Cone and cone-scale with adhering wing. Fig. 139, Seed and wing. Fig. 140, Leaf-fascicle and magnified leaf-section. Fig. 141, Parti-colored bark. Fig. 142, Tree with white trunk.

## 17. PINUS GERARDIANA

- 1832 P. Gerardiana Wallich ex Lambert, Gen. Pin. ed. 8vo, ii. t. 79.

Spring-shoots glabrous. Leaves from 6 to 10 cm. long, serrulate; stomata dorsal and ventral; resin-ducts external. Scales of the conelet armed with a short spine. Cones from 9 to 15 cm. long, short-pedunculate, ovoid or oblong; apophyses fulvous brown, very thick, with a prominent reflexed or erect protuberance culminating in an umbo on which the spine is more or less persistent; nuts remarkably long, narrow, terete, the shell fragile, the short wing falling with the nut or adhering to the adjacent scale.

A tree of the northwestern Himalayas found on the borders of Cashmere and Thibet and in Kafiristan and north Afghanistan, and so highly prized for its nuts that it is rarely felled for its wood. It grows in dry regions and rarely attains a height of 20 metres. Attempts to cultivate this species, even in the milder parts of Great Britain, have generally failed.

The apophysis of the cone varies much in prominence (figs. 134, 135), but the peculiar seed is invariable and quite unlike that of any other Pine. The general color of the trunk at a distance is silver-gray.

Plate XIV.

Fig. 133, Cone. Fig. 134, Cone-scale with adhering seed-wing. Fig. 135, Cone-scale of flatter form. Fig. 136, Seed and wing. Fig. 137, Leaf-fascicle and magnified leaf-section.

41

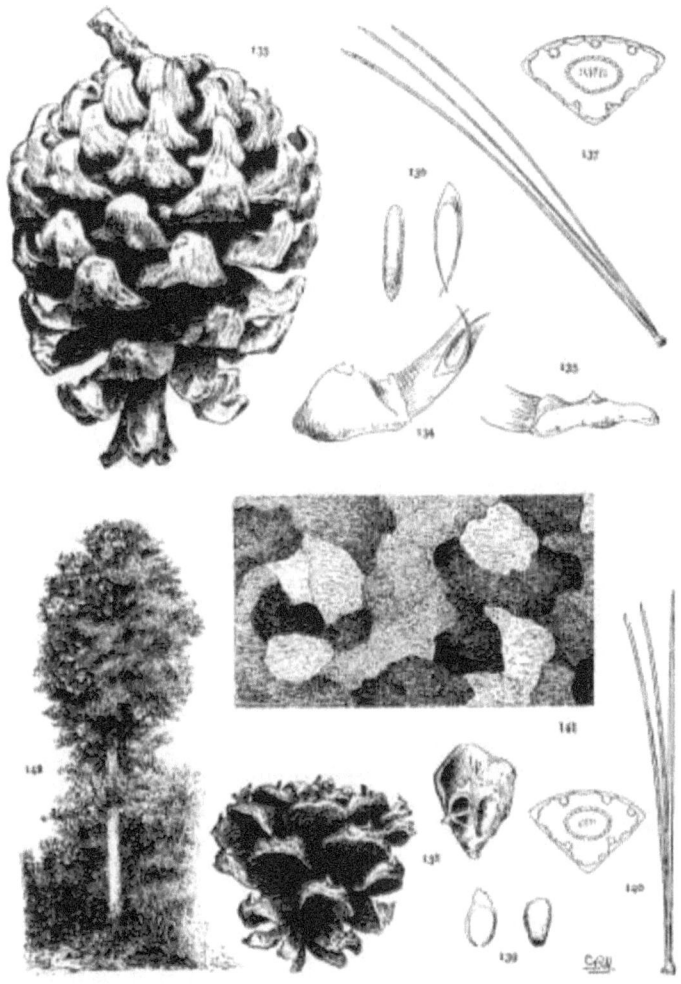

PLATE XIV. P. GERARDIANA (133-137), BUNGEANA (138-142)

## VI. BALFOURIANAE

Seeds with long effective wings. Leaves entire, in fascicles of 5, the sheath deciduous.

The two species known as Foxtail Pines are alike in their short entire falcate leaves, persisting for many years and forming long dense foliage-masses. They differ in the armature of their cones and in their seed-wings. The presence of both adnate and articulate wings in these closely related species suggests that these two forms of wing are not fundamentally distinct.

Cone-scales short-mucronate, the seed-wing adnate 18. Balfouriana.

Cone-scales long-aristate, the seed-wing articulate    19. aristata.

## 18. PINUS BALFOURIANA

- 1853 P. Balfouriana Balfour in Bot. Exp. Oregon, 1, f.

Spring-shoots somewhat puberulent. Leaves from 2 to 4 cm. long, persistent for many years; stomata ventral only; resin-ducts external. Scales of the conelet short-mucronate. Cones from 7 to 12 cm. long, tapering to a rounded apex, short-pedunculate; apophyses dark terracotta-brown, tumid, the umbo bearing a short recumbent prickle; seed with a long adnate wing.

44

An alpine species growing often at the timber-limit. It is found in two distinct stations in California, on the northern Coast Range and on the southern Sierras. It is not often cultivated, but young plants may be seen in the Arnold Arboretum and in the Royal Gardens at Kew.

Plate XV.

Fig. 147, Cone, seed and enlarged cone-scale. Fig. 148, Leaf-fascicle. Fig. 149, Magnified leaf-section. Fig. 150, A branch with persistent leaves.

## 19. PINUS ARISTATA

- 1862 P. aristata Engelmann in Am. Jour. Sci. ser. 2, xxxiv. 331.

- 1871 P. Balfouriana Watson in King's Rep. v. 331 (not Balfour).

Spring-shoots glabrous or temporarily pubescent. Leaves from 2 to 4 cm. long, persistent for many years; stomata ventral only; resin-ducts external. Scales of the conelet prolonged into long slender bristles. Cones from 4 to 9 cm. long, subcylindrical or tapering to a rounded apex, short-pedunculate; apophyses terracotta or purple-brown, tumid, the long bristles of the umbo often partly or wholly broken away; seeds with a long articulate wing.

A bushy tree, similar in foliage to the preceding species, growing at the timber-limit from Colorado through Utah, central and southern Nevada and northern Arizona into southeastern California, but separated from the nearest station of P. Balfouriana by an arid treeless desert. Engelmann (in Brewer and Watson, Bot. Calif. ii. 125) considered it to be a variety of P. Balfouriana.

Plate XV.

Fig. 143, Cone. Fig. 144, Seed and enlarged cone-scale. Fig. 145, Leaf-fascicle and magnified leaf-section. Fig. 146, Conelet.

PLATE XV. P. ARISTATA (143-146), BALFOURIANA (147-150)

# DIPLOXYLON

Bases of the bracts subtending leaf-fascicles decurrent. Leaves serrulate; fibro-vascular bundle double; stomata dorsal and ventral.

Cones with a dorsal umbo, the phyllotaxis complex. Wood hard, with dark resinous bands, the annual rings clearly defined.

In this section there are a few species combining the essential characters of Diploxylon with important characters of Haploxylon. A subsection, Parapinaster, is established for these exceptional species.

c. Parapinaster — Species with the fascicle-sheath or seed-wing of Haploxylon.

d. Pinaster — Sheath persistent, seed-wing articulate, effective.

**Parapinaster**

| | |
|---|---|
| Sheath of the leaf-fascicle deciduous | VII. Leiophyllae. |
| Sheath of the leaf-fascicle persistent. | |
| Seed-wing of the Strobi | VIII. Longifoliae. |
| Seed-wing of the Gerardianae | IX. Pineae. |

**VII. LEIOPHYLLAE**

Sheath of the leaf-fascicles deciduous.

| | |
|---|---|
| Leaves short, erect, the fructification triennial | 20. leiophylla. |
| Leaves long, pendent, the fructification biennial | 21. Lumholtzii. |

**20. PINUS LEIOPHYLLA**

- 1831 P. leiophylla Schlechtendal and Chamisso in Linnaea, vi. 354.
- 1848 P. chihuahuana Engelmann in Wislizenus, Tour. Mex. 103.

46

Spring-shoots uninodal. Leaves in fascicles of 3, 4 or 5, the sheath deciduous, from 8 to 14 cm. long; resin-ducts medial with an occasional internal duct. Conelets single or verticillate, their scales mucronate; conelets of the second year only slightly enlarged. Cones maturing the third year, not exceeding 7 cm. in length, ovate or ovate-conic, subsymmetrical, more or less reflexed, persistent for several years on some trees, sometimes serotinous; apophyses lighter or darker brown, often with an olive or fuscous shade, thin or tumid, the umbo double, the mucro more persistent near the apex of the cone.

This species grows at subtropical or warm-temperate altitudes in Mexico, from Oaxaca through the central and western states to southern Arizona and New Mexico. As it approaches the northern part of its range the leaves become thicker and more rigid and the number in the fascicle is reduced to 3 or 4 (var. chihuahuana, Shaw, Pines Mex. 14). Like P. rigida it sprouts freely along the branches and trunk, and stumps of felled trees put out shoots in great numbers. The species is easily recognized by the deciduous sheath and triennial cone.

Plate XVI.

Fig. 151, Branch with fruit of first, second and third years. Fig. 152, Leaf-fascicles. Fig. 153, Magnified leaf-section of the species. Fig. 154, Magnified leaf-section of the variety.

## 21. PINUS LUMHOLTZII

- 1894 P. Lumholtzii Robinson & Fernald in Proc. Am. Acad. xxx. 122.

Spring-shoots uninodal, sometimes multinodal. Leaves in fascicles of 3, the sheath deciduous, from 20 to 30 cm. long, absolutely pendent; resin-ducts medial and internal. Conelets subterminal, or lateral and subterminal, mucronate. Cones not exceeding 7 cm. in length, symmetrical, pendent on slender peduncles, ovate-conic, early deciduous; apophyses sublustrous, nut-brown, tumid at the margins, flat on the surface, the umbo large, the mucro rarely persistent.

A remarkable Pine with long pendent bright green foliage, confined to the western states of Mexico and ranging on the mountains from southern Jalisco to the latitude of the city of Chihuahua. Each season's growth of leaves hangs from the branchlet like a long beard, from which the tree receives, in some localities, the name

"Pino barba caida." In the herbarium the long leaves, deciduous sheaths, and the decurrent bases of the bracts, present a combination of characters not found in other species.

Plate XVI.

Fig. 155, Cone. Fig. 156, Cone. Fig. 157, Leaf-fascicle. Fig. 158, Magnified leaf-section. Fig. 159, Tree at Ferraria de Tula.

PLATE XVI. P. LEIOPHYLLA (151-154), LUMHOLTZII (155-159)

## VIII. LONGIFOLIAE

Seed-wing adnate to the nut. Leaves long, in fascicles of 3, the sheath persistent.

Apophysis of the cone prolonged and reflexed    22. longifolia.

Apophysis of the cone low-pyramidal     23. canariensis.

## 22. PINUS LONGIFOLIA

- 1803 P. longifolia Roxburgh ex Lambert, Gen. Pin. i. 29, t. 21.
- 1897 P. Roxburghii Sargent, Silva N. Am. xi. 9.

Spring-shoots uninodal. Leaves in fascicles of 3, the sheath persistent, from 20 to 30 cm. long; resin-ducts external, the hypoderm often in large masses, some or all of the endoderm cells with thick outer walls. Cones from 10 to 17 cm. long, short-pedunculate, ovoid-conic; apophyses lustrous brown-ochre or fuscous brown, elevated into thick, often reflexed, beaks with obtuse mutic umbos; seeds with large nuts and adnate striated dark gray or fuscous brown wings.

48

Of the three Pines of the Himalayas this species is the most important. It grows on the outer slopes and foot-hills from Bhotan to Afghanistan. The wood is used for construction and for the manufacture of charcoal, the thick soft bark is valuable for tanning, the resin is abundant and of commercial importance, and the nuts are gathered for food. The tree is not hardy in cool-temperate climates, but has been successfully grown in northern Italy.

It differs from P. canariensis in the usually protuberant apophysis of the cone, in the thick outer walls of the leaf-endoderm and in the nearly smooth walls of the ray-tracheids of the wood. In the dimensions of cone and leaf, in the dermal tissues and resin-ducts of the leaf and in the peculiar coloring of the seed-wing, the two species are alike.

Plate XVII.

Fig. 160, Cone. Fig. 161, Leaf-fascicle. Fig. 162, Magnified leaf-section.

## 23. PINUS CANARIENSIS

- 1825 P. canariensis Smith in Buch, Canar. Ins. 159.

Spring-shoots uninodal, pruinose. Bud-scales with conspicuously long free fimbriate margins. Leaves in fascicles of 3, the sheath persistent, from 20 to 30 cm. long; the hypoderm often in large masses, the resin-ducts external, the endoderm with thin outer walls. Cones from 10 to 17 cm. long, short-pedunculate, ovoid-conic; apophyses lustrous or sublustrous nut-brown, more or less pyramidal, the umbo unarmed; seeds as in the last species.

A species confined to the Canary Islands, but cultivated in northern Italy. The stately habit of this tree is seen in Schröter's portrait (Exc. Canar. Ins. t. 15).

Plate XVII.

Fig. 163, Cone and seed. Fig. 164, Magnified leaf-section. Fig. 165, Habit of the tree.

47

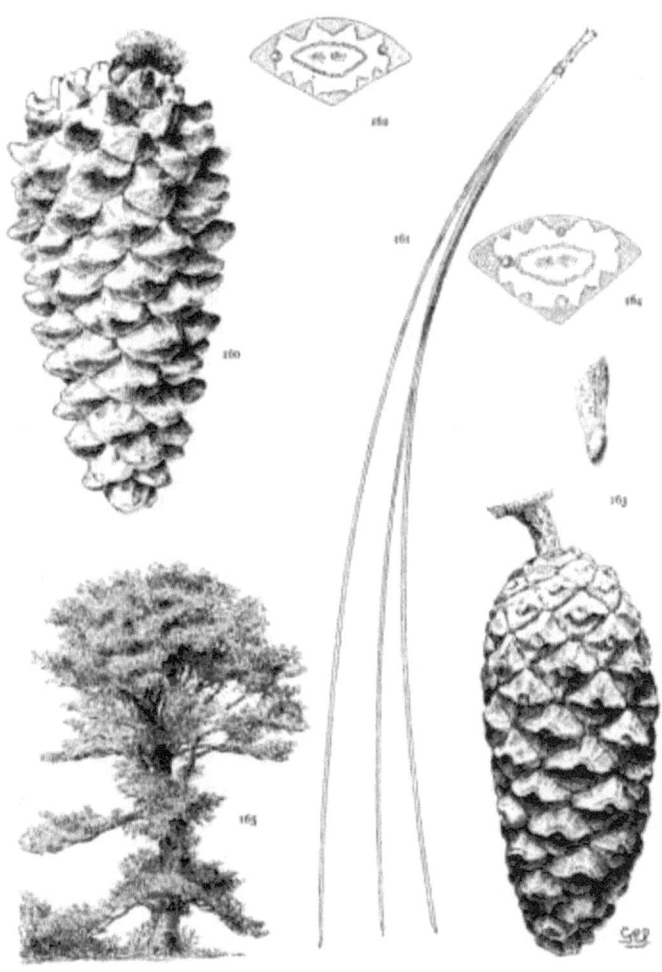

PLATE XVII. P. LONGIFOLIA (160-162), CANARIENSIS (163-165)

## IX. PINEAE

Seed-wing articulate, short, ineffective. Leaves binate, the sheath persistent. One species only.

## 24. PINUS PINEA

- 1753 P. pinea Linnaeus, Sp. Pl. 1000.
- 1778 P. sativa Lamarck, Fl. Franç. ii. 200.
- 1854 P. maderiensis Tenore in Ann. Sci. Nat. sér. 4, ii. 379.

Spring-shoots uninodal. Leaves from 12 to 20 cm. long; resin-ducts external. Conelet mutic, slightly larger in the second year. Cones triennial, from 10 to 14 cm. long, ovoid or subglobose; apophyses lustrous nut-brown, convex, of large size, the umbo double; seeds large with a short, loosely articulated, deciduous wing.

A species of the Mediterranean Basin, from Portugal to Syria. Its northern limit is in southern France and northern Italy, but it is cultivated in the southern parts of the British Isles and is a familiar ornament of park and garden in southern Europe, and is valued for its peculiar beauty and for its large savory nuts. In wood anatomy as well as in the seed it agrees with the Gerardianae of the Soft Pines.

Plate XVIII.

Fig. 166, Fruit of three seasons. Fig. 167, Cone-scales and seed. Fig. 168, Magnified leaf-section. Fig. 169, Habit of the tree.

49

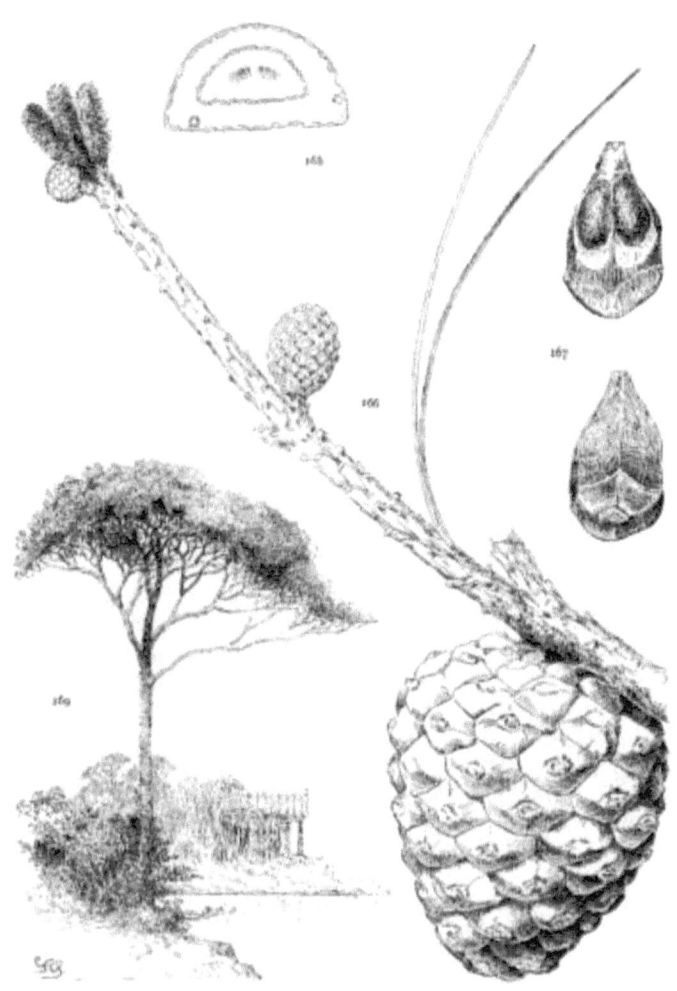

PLATE XVIII. PINUS PINEA

50

**Pinaster**

Bases of the bracts subtending leaf-fascicles decurrent. Seeds with an effective articulate wing. Umbo of the cone-scales dorsal. Leaves serrulate, stomatiferous on all faces, the sheath persistent. Walls of the tracheids of the medullary rays dentate.

Forty-two of the sixty-six species of Pinus are included in this subsection. As a group they are clearly circumscribed by several correlated characters and are more closely interrelated than the twenty-four species previously described. The distinctions of umbo and seed have disappeared. The umbo here is invariably dorsal, the seed-wing invariably articulate.

New forms, however, are gradually evolved — the seed with a thick wing-blade, the indurated oblique cone, the serotinous cone with its intermittent seed-release, and the multinodal spring-shoot. There are, moreover, new forms of leaf-hypoderm and a new position of the resin-duct.

Of these new characters, the thick wing-blade attains such proportions in the three species of the Macrocarpae that they can be grouped apart. But the characters that finally culminate in a lateral oblique serotinous cone are so gradually and irregularly developed that they offer no divisional distinctions. With the aid of wood and leaf characters, however, groups can be established which preserve the evolutionary sequence and, at the same time, the obvious affinity of the species.

Wing-blade thin or slightly thickened at the base.
Cones dehiscent at maturity.

| | |
|---|---|
| Pits of the ray-cells large | X. Lariciones |
| Pits of the ray-cells small | XI. Australes |
| Cones serotinous, pits of the ray-cells small | XII. Insignes |
| Wing-blade very thick | XIII. Macrocarpae |

The species of this subsection are very difficult, if not impossible, to classify by the usual method, which groups all species under a few characters assumed to be invariable and of fundamental importance. Such a method can be successfully applied to the Soft Pines and to some of the Hard Pines, but cannot be applied to all the Hard Pines without forcing some of them into unnatural associations.

To take an example, the group Pseudostrobus, characterized by pentamerous leaf-fascicles, appears in many systems. In this group are placed P. Torreyana and P. leiophylla. Another group, with trimerous fascicles, contains P. Sabiniana and P. taeda. Now there are no two species more obviously related by important peculiarities than P. Torreyana and P. Sabiniana; nevertheless they are, by this method, kept apart and associated with species which they resemble in no important particular.

An attempt is made here to avoid such incongruities. Groups X, XI and XII represent different stages of evolution. In the Lariciones the cone is symmetrical, and dehiscent and deciduous at maturity, while the spring-shoot is uninodal. In the Australes there is a similar cone, but the spring-shoot gradually becomes multinodal. In the Insignes the cone is oblique, persistent and serotinous, and the spring-shoot is multinodal.

These definitions state the degree of evolution attained by each group, but not all the species of a group conform exactly with its definition. In each group are species with a characteristic of another group. Among the Lariciones are a few species with both symmetrical and oblique cones, and two with persistent cones. Similar exceptions occur among the Australes. Among the Insignes are a few species with symmetrical cones, and two with cones that are rarely, if ever, serotinous.

There is, however, no difficulty in fixing the systematic position of these exceptional species through other characters which show their true affinity. They are placed with the species which they most resemble. Their exceptional characters are merely the evidence of the evolution that pervades and unites the groups. Therefore the definition of a group is not necessarily the exact definition of its species, and a species is placed in a group because all its characters, specific and evolutional, show a closer affinity with that group than with the species of any other.

## X. LARICIONES

Pits of the ray-cells large. Cells of the leaf-hypoderm uniform. Spring-shoots uninodal. Cones dehiscent at maturity.

This group represents the first stage in the evolution of the Hard Pines. All the species, like the Soft Pines, are uninodal and the cones are dehiscent at maturity, but the trend toward the serotinous species is shown in the occasional appearance of the oblique cone as a varietal form of a few species, and in the persistent cone of the last two species of this group.

All the species of this group are of the Old World except P. resinosa and P. tropicalis. These two are the only American Pines combining large pits with dentate tracheids, and are the only American Hard Pines with external resin-ducts of the leaf.

| | |
|---|---|
| Cones deciduous at maturity. | |
| Cones ovate or ovate-conic. | |
| Conelet with tuberculate or entire scales. | |
| Resin-ducts external and medial | 25. resinosa |
| Resin-ducts septal and external | 26. tropicalis |
| Conelet with mucronate scales. | |
| Resin-ducts mostly external. | |
| Conelet pedunculate, erect. | |
| Cone nut-brown | 27. Massoniana |
| Cone dull tawny yellow | 28. densiflora |
| Conelet pedunculate, reflexed | 29. sylvestris |
| Conelet subsessile, erect | 30. montana |
| Resin-ducts mostly medial. | |
| Bark-formation late | 31. luchuensis |
| Bark-formation early. | |
| Cone nut-brown | 32. Thunbergii |
| Cone lustrous tawny yellow | 33. nigra |
| Cones narrow cylindrical | 34. Merkusii |
| Cones tenaciously persistent. | |
| Leaves stout, relatively short | 35. sinensis |
| Leaves slender, relatively long | 36. insularis |

## 25. PINUS RESINOSA

- 1789 P. resinosa Aiton, Hort. Kew. iii. 367.
- 1810 P. rubra Michaux f. Hist. Arbr. Am. i. 45, t. 1.

Spring-shoots uninodal. Leaves binate, from 12 to 17 cm. long; resin-ducts external or external and medial; hypoderm uniform and inconspicuous. Scales of the conelet mutic. Cones from 4 to 6 cm. long, subsessile, symmetrical, deciduous the third year, leaving a few basal scales on the tree; apophyses sublustrous, nut-brown, somewhat thickened along a transverse keel.

From Nova Scotia and Lake St. John this species ranges westward to the Winnipeg River and southward into Minnesota, Michigan, northern New York and eastern Massachusetts, with rare occurrence on the mountains of Pennsylvania. Under cultivation it is a beautiful tree, adapted to cold-temperate climates. It was considered by Loiseleur (1812) and by Spach (1842) to be a variety of P. nigra (laricio). The two species vary in the color of the cone, the anatomy of the leaves, the buds, and in the armature of the conelet. A fallen cone of this species is moreover usually imperfect from the loss of a few basal scales.

Plate XIX.

Fig. 170, Cone and enlarged conelet. Fig. 171, Leaf-fascicle and magnified leaf-section.

## 26. PINUS TROPICALIS

- 1851 P. tropicalis Morelet in Rev. Hort. Côte d'Or, i. 105.
- 1904 P. terthrocarpa Shaw in Gard. Chron. ser. 3, xxxv. 179, f. 74.

Spring-shoots uninodal. Leaves binate, sometimes ternate, from 15 to 30 cm. long, rigid, erect; hypoderm of uniform thick-walled cells; resin-ducts of remarkable size, septal, or not quite touching the endoderm and technically external. Scales of the conelet minutely tuberculate. Cones from 5 to 8 cm. long, short-pedunculate, erect or patulous; ovate-conic, symmetrical; apophyses rufous brown, low-pyramidal, the umbo mutic.

Growing at sea-level within the tropics and confined to western Cuba and the Isle of Pines. On the island it is associated with P. caribaea. This species needs no other means of identification than its peculiar leaf-section. Septal ducts are found in P. oocarpa, Pringlei, Merkusii and rarely in other species, but they never attain the extraordinary size that appears to be invariable in P. tropicalis.

Plate XIX.

Fig. 172, Cone and enlarged conelet. Fig. 173, Branch with leaves, much reduced. Fig. 174, Leaf-fascicle and magnified leaf-section. Fig. 175, Trees on the Isle of Pines.

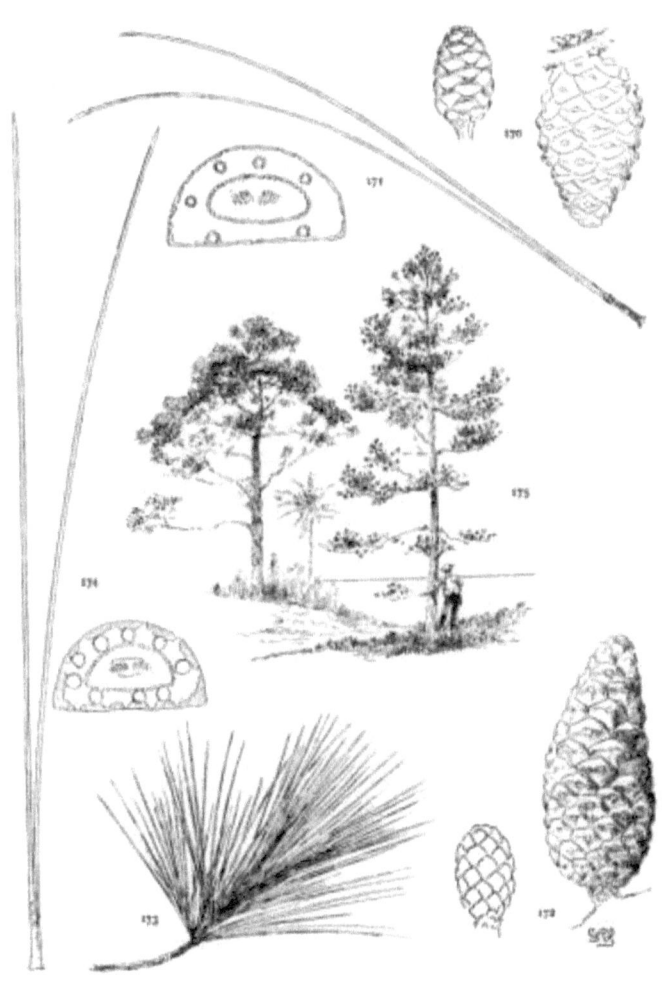

PLATE XIX. P. RESINOSA (170, 171), TROPICALIS (172-175)

## 27. PINUS MASSONIANA

- 1803 P. Massoniana Lambert, Gen. Pin. i. 17, t. 12. 1861 P. canaliculata Miquel in Jour. Bot. Neerland. i. 86.

Spring-shoots uninodal. Leaves binate, rarely ternate, from 12 to 20 cm. long, slender and pliant; hypoderm inconspicuous; resin-ducts external. Staminate catkins often in long dense clusters. Conelets partly tuberculate or mucronate, partly mutic. Cones symmetrical, from 4 to 7 cm. long, ovate-conic, short-pedunculate, early deciduous; apophyses sublustrous, nut-brown, flat or somewhat elevated, the umbo usually mutic.

The Chinese Red Pine is found in warm-temperate climates. It is native to southeastern China and follows the valley of the Yangtse River into Szech'uan. It has been confused by London with P. pinaster, which it resembles in no respect, by Siebold with P. Thunbergii, from which it differs in leaf-dimensions and in leaf-section, and by Mayr with his P. luchuensis, whose peculiar cortex and whose leaf-section has no counterpart among Chinese Hard Pines. Its nearest relative is P. densiflora, from which it differs in its longer leaves, in the color of its cone and in its conelet (Plate XX, figs. 176, 179).

Plate XX.

Fig. 176, Cone and enlarged conelet. Fig. 177, Two leaf-fascicles. Fig. 178, Magnified leaf-section.

## 28. PINUS DENSIFLORA

- 1842 P. densiflora Siebold & Zuccarini, Fl. Jap. ii. 22, t. 112.
- 1854 P. scopifera Miquel in Zollinger, Syst. Verz. Ind. Archip. 82.

Spring-shoots more or less pruinose, uninodal. Leaves binate, from 8 to 12 cm. long, slender; hypoderm of few inconspicuous cells; resin-ducts external. Staminate catkins in long dense clusters. Scales of the conelet conspicuously mucronate. Cones symmetrical, from 3 to 5 cm. long, ovate-conic, often persistent for a few years but with a weak hold on the branch; apophyses dull pale tawny yellow, flat or slightly elevated, the mucro more or less persistent.

The Japanese Red Pine forms extensive forests on the mountains of central Japan. It is perfectly hardy in cold-temperate climates. Wild specimens of China, ascribed to this species, are forms of the variable P. sinensis. From P. Massoniana it differs in its shorter leaves and yellow cone, but particularly in the more prominent prickles and thicker scales of its conelet (figs. 176, 179).

Plate XX.

Fig. 179, Cones and enlarged conelet. Fig. 180, Leaf-fascicles. Fig. 181, Magnified leaf-section and more magnified dermal tissues of the leaf.

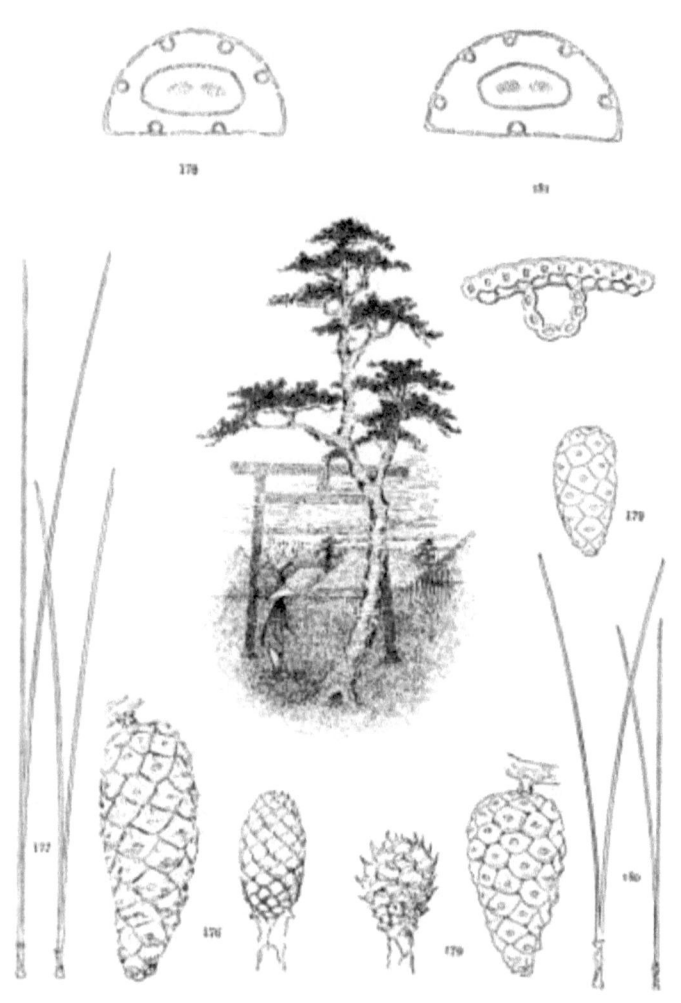

PLATE XX. P. MASSONIANA (176-178), DENSIFLORA (179-181)

## 29. PINUS SYLVESTRIS

- 1753 P. sylvestris Linnaeus, Sp. Pl. 1000 (excl. var.).
- 1768 P. rubra Miller, Gard. Dict. ed. 8.
- 1768 P. tatarica Miller, Gard. Dict. ed. 8.

- 1781 P. mughus Jacquin, Icon. Pl. Rar. i. t. 193 (not Scopoli).
- 1798 P. resinosa Savi, Fl. Pisa. ii. 354 (not Aiton).
- 1827 P. humilis Link in Abhandl. Akad. Berlin, 171.
- 1849 P. Kochiana Klotzsch in Linnaea, xxii. 296.
- 1849 P. armena Koch in Linnaea, xxii. 297.
- 1849 P. pontica Koch in Linnaea, xxii. 297.
- 1859 P. Frieseana Wichura in Flora, xlii. 409.
- 1906 P. lapponica Mayr, Fremdl. Wald- & Parkb. 348.

Spring-shoots uninodal. Leaves binate, from 3 to 7 cm. long; hypoderm inconspicuous; resin-ducts external. Conelet reflexed, minutely mucronate. Cones from 3 to 6 cm. long, reflexed, symmetrical or sometimes oblique, ovate-conic, deciduous; apophyses dull pale tawny yellow of a gray or greenish shade, flat, elevated or protuberant and often much more prominent on the posterior face of the cone, the umbo with a minute prickle or its remnant.

A tree of great commercial value, with a very extended range, from Norway, Scotland and southern Spain to northeastern Siberia. A vigorous hardy species and extensively cultivated. The red upper trunk, characteristic of this Pine, is not invariable. The dark upper trunk is sufficiently common to be considered a varietal form (Mathieu, Flore Forest. ed. 4, 582). In various localities may be found trees bearing oblique cones, their apophyses showing various degrees of protuberance up to the extreme development represented in Loudon's illustration of the variety uncinata (Arb. Brit. iv, f. 2047). This cone is the beginning of the changes that culminate in species with oblique cones only. In P. sylvestris, however, the purpose of this form of cone is not apparent except in connection with this evolution.

Plate XXI.

Figs. 182, 183, Cones. Fig. 184, Leaf-fascicle, magnified leaf-section and more magnified dermal tissues of the leaf. Fig. 185, Habit of the tree.

## 30. PINUS MONTANA

- 1768 P. montana Miller, Gard. Dict. ed. 8.
- 1772 P. mughus Scopoli, Fl. Carn. ii. 247.
- 1791 P. pumilio Haenke in Jirasek, Beobacht. 68.
- 1804 P. mugho Poiret in Lamarck, Encycl. Méth. v. 336.
- 1805 P. uncinata Ramond ex De Candolle, Lamarck, Fl. Franç. ed. 3, iii. 726.
- 1813 P. sanguinea Lapeyrouse, Hist. Pl. Pyren. 587.
- 1827 P. rotundata Link in Abhandl. Akad. Berlin, 168.
- 1830 P. obliqua Sauter ex Reichenbach, Fl. Germ. Exc. 159.
- 1837 P. uliginosa Neumann ex Wimmer, Arb. Schles. Ges. 95.

Spring-shoots uninodal. Leaves binate, from 3 to 8 cm. long, the epiderm very thick, hypoderm weak; resin-ducts external. Conelets mucronate, nearly sessile. Cones from 2 to 7 cm. long, subsessile, ovate or ovate-conic, symmetrical or oblique, often persistent; apophyses lustrous tawny-yellow or dark brown, both colors often shading into each other on the same cone, flat, prominent or prolonged 56 into uncinate beaks of various lengths, the last much more developed on the posterior face of the cone, the umbo bordered by a narrow dark ring and bearing the remnant of the mucro.

P. montana grows as a bush or as a small tree, the two forms often associated. It ranges from central Spain through the Pyrenees, Alps and Apennines to the Balkan Mountains, associated with P. cembra at higher, with P. sylvestris at lower altitudes. It grows indifferently in bogs and on rocky slopes. Its dwarf form, under the name of the Mugho Pine, is extensively cultivated as a garden ornament.

On the differences of the cone this species has been divided into three subspecies: uncinata, with an oblique cone and protuberant apophyses; pumilio, with a symmetrical cone and an excentric umbo; mughus, with a symmetrical cone and a concentric umbo. Other segregations based on the degree of development of the apophysis and on the size and color of the cone, have received names of four or even five terms—Pinus montana pumilio applanata—or Pinus montana uncinata rostrata castanea etc., etc. These elaborations may be seen in the Tharand Jahrbuch of 1861, p. 166, and with them appear also Hartig's specifications of 60 forms of this species, each dignified with a Latin name.

Plate XXI.

Fig. 186, Cone of var. uncinata. Figs. 187, 188, Cones. Fig. 189, Leaf-fascicles, magnified leaf-section and more magnified dermal tissues of the leaf. Fig. 190, Tree and dwarf-form of the Pyrenees.

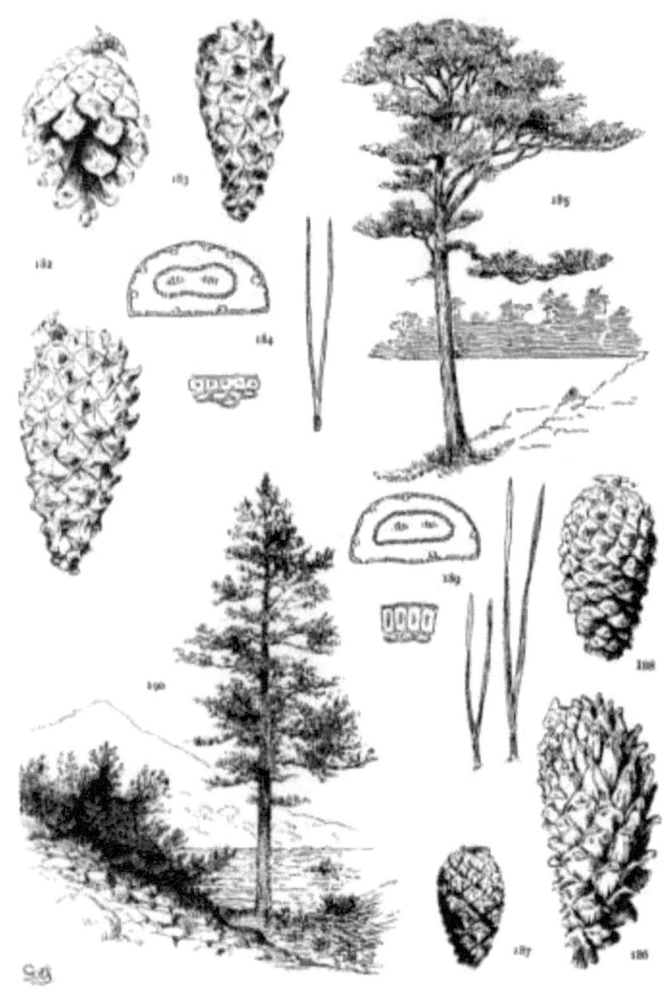

PLATE XXI. P. SYLVESTRIS (182-185), MONTANA (186-190)

## 31. PINUS LUCHUENSIS

- 1894 P. luchuensis Mayr in Bot. Centralbl. lviii. 149, f.

Spring-shoots uninodal. Bark-formation late, the upper trunk covered with a smooth cortex. Leaves binate, from 12 to 16 cm. long, the epiderm thick, hypoderm of two or three rows of cells; resin-ducts medial or with an occasional external duct. Conelets mucronate toward the apex. Cones from 3 to 6 cm. long, ovate-conic, symmetrical; apophyses lustrous nut-brown, transversely carinate, the umbo unarmed.

This Pine is known to me through Mayr's description and a single dried specimen. The smooth cortex of young trees distinguishes it from all other east-Asiatic Hard Pines. Mayr includes under this species the Pine of Hong Kong. But in this he must be mistaken, for there is no species yet found in China that agrees with the description of P. luchuensis.

Plate XXII.

Fig. 191, Cone. Fig. 192, Leaf-fascicle and magnified leaf-section.

## 32. PINUS THUNBERGII

- 1784 P. sylvestris Thunberg, Fl. Jap. 274 (not Linnaeus).
- 1842 P. Massoniana Siebold & Zuccarini. Fl. Jap. ii. 24, t. 113 (not Lambert).
- 1868 P. Thunbergii Parlatore in DC. Prodr. xvi-2, 388.

Spring-shoots uninodal. Buds of leading-shoots white and conspicuous. Leaves binate, from 6 to 11 cm. long, the epiderm thick, hypoderm strong, resin-ducts medial. Conelets with short-mucronate scales. Cones from 4 to 6 cm. long, ovate or ovate-conic, symmetrical; apophyses nut-brown, flat or convex and transversely carinate, the prickle of the umbo more or less persistent.

The Black Pine of Japan has been cultivated for centuries, and by skillful Japanese gardeners has been trained into dwarf and other curious forms. It is hardy in cold-temperate climates. It is distinct from P. densiflora by the medial ducts of its leaf, from P. nigra by the fewer, larger, brown scales of its cone, and from P. resinosa by the armature of its conelet. It appears in most determinations of

Chinese collections, but there is no Chinese Pine with the white buds and the medial leaf-ducts of this species.

Plate XXII.

Fig. 196, Two cones. Fig. 197, Leaf-fascicle and magnified leaf-section.

58

## 33. PINUS NIGRA

- 1785 P. nigra Arnold, Reise n. Mariaz. 8, t.
- 1804 P. laricio Poiret in Lamarck, Encycl. Méth. v. 339.
- 1808 P. halepensis Bieberstein, Fl. Taur. Cauc. ii. 408 (not Miller).
- 1809 P. pinaster Besser, Fl. Galic. ii. 294 (not Aiton).
- 1813 P. maritima Aiton, f. Hort. Kew. v. 315 (not Lambert).
- 1816 P. sylvestris Baumgarten, Stirp. Transsilv. ii. 304 (not Linnaeus).
- 1818 P. pyrenaica Lapeyrouse, Hist. Pl. Pyren. Suppl. 146.
- 1824 P. Pallasiana Lambert, Gen. Pin. ii. 1, t. 1.
- 1825 P. austriaca Höss in Flora, viii-1, Beil. 113.
- 1831 P. nigricans Host, Fl. Austr. ii. 628.
- 1842 P. dalmatica Visiani, Fl. Dalmal. 199, note.
- 1851 P. Salzmanni Dunal in Mém. Acad. Montp. ii. 82, tt.
- 1863 P. Heldreichii Christ in Verh. Nat. Ges. Basel, iii. 549.
- 1864 P. leucodermis Antoine in Oesterr. Bot. Zeitschr. xiv. 366.
- 1896 P. pindica Formanek in Verh. Nat. Ver. Brünn, xxxiv. 272.

Spring-shoots uninodal. Leaves binate, from 9 to 16 cm. long, the epiderm thick, hypoderm conspicuous, resin-ducts medial. Conelets mucronate. Cones from 4 to 8 cm. long, subsessile, symmetrical; apophyses lustrous, tawny yellow, transversely carinate, the keel strongly convex, the mucro of the umbo more or less persistent.

A valuable tree unequally distributed over the mountain slopes of central and southern Europe and Asia Minor. The typical form, under the name of the Austrian Pine, is a familiar exotic of the Middle and Eastern States of America. As Mathieu states (Flore Forest., ed. 4, 597), this species is quite constant in cone and bark. It may be added that the anatomy of the leaf is also constant, while the dimensions of both leaf and cone present no unusual variations. The varieties generally accepted are founded on the habit of the tree, a character of forestal or horticultural rather than of botanical importance.

Plate XXII.

Fig. 193, Two cones. Fig. 194, Leaf-fascicle and magnified leaf-section. Fig. 195, Magnified dermal tissues of the leaf.

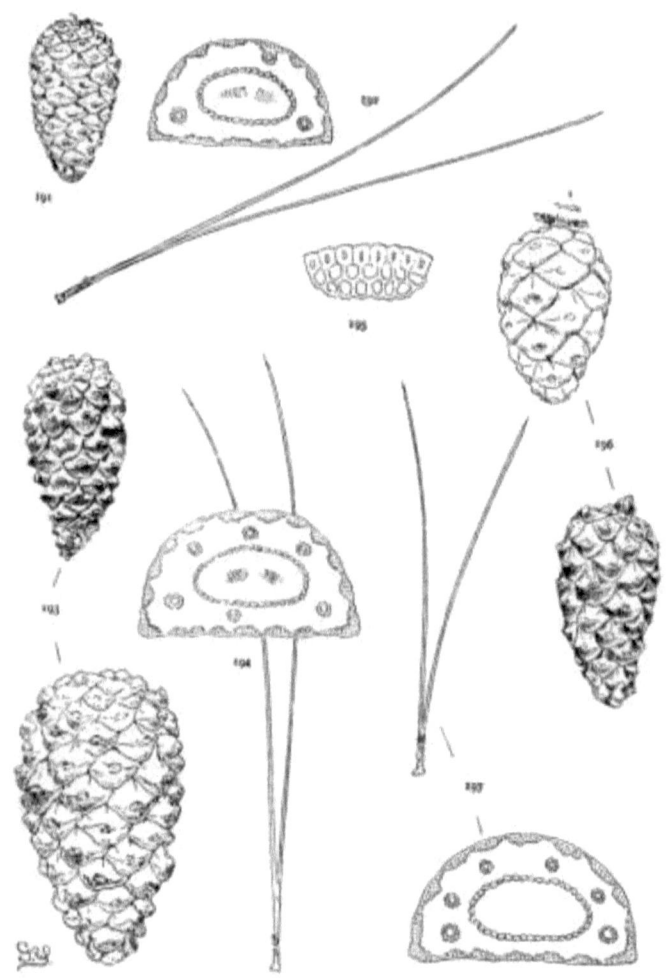

PLATE XXII. P. LUCHUENSIS (191, 192), NIGRA (193-195), THUNBERGII (196, 197)

## 34. PINUS MERKUSII

- 1790 P. sylvestris Loureiro, Fl. Cochinch. ii. 579 (not Linnaeus).

- 1845 P. Merkusii De Vriese, Pl. Nov. Ind. Bat. 5, t. 2.
- 1847 P. Finlaysoniana Wallich ex Blume, Rumphia, iii. 210.
- 1849 P. Latteri Mason in Jour. Asiat. Soc. i. 74.

Spring-shoots uninodal. Leaves binate, slender, from 15 to 20 cm. long, the hypoderm of uniform thick-walled cells, resin-ducts medial, or with internal or septal ducts, endoderm-cells very unequal in size, some of them large. Conelets unarmed. Cones from 5 to 8 cm. long, peculiarly narrow-cylindrical, symmetrical; apophyses lustrous, rufous brown, radially carinate, the transverse keel prominent.

Of the habit of this Pine I know nothing. As a species it is very clearly defined by its peculiar cone and leaf-section. It grows in the Philippines, Sumatra, Lower Burmah and western Indo-China. In my specimen the pits of the ray-cells of the wood are both large and small. In this particular it may belong in either of two groups of species. Its uniform leaf-hypoderm associates it with this group or with P. halepensis of the Insignes. I have assumed the cone to be dehiscent at maturity and have placed it with the Lariciones, but if further information shows the cone to be serotinous, this species should be transferred to the serotinous group.

Plate XXIII.

Fig. 198, Cone. Fig. 199, Magnified sections of two leaves. Fig. 200, Leaf-fascicle.

## 35. PINUS SINENSIS

- 1832 P. sinensis Lambert, Gen. Pin. ed. 8vo. i. 47, t. 29.
- 1867 P. tabulaeformis Carrière, Trait. Conif. ed. 2, 510.
- 1881 P. leucosperma Maximowicz in Bull. Acad. St. Pétersb. xxvii. 558.
- 1899 P. yunnanensis Franchet in Jour. de Bot. xiii. 253.
- 1901 P. funebris Komarow in Act. Hort. Petrop. xx. 177.
- 1902 P. Henryi Masters in Jour. Linn. Soc. xxvi. 550.
- 1906 P. densata Masters in Jour. Linn. Soc. xxxvii. 416.

- 1906 P. prominens Masters in Jour. Linn. Soc. xxxvii. 417.
- 1911 P. Wilsonii Shaw in Sargent, Pl. Wilson. i. 3.

Spring-shoots uninodal, pruinose. Leaves binate, ternate, or both, from 10 to 15 cm. long, stout and rigid; resin-ducts external, or external and medial. Staminate catkins in short capitate clusters. Conelets mucronate. Cones from 4 to 9 cm. long, ovate, symmetrical or oblique, tenaciously persistent, dehiscent at maturity; apophyses lustrous, pale tawny yellow at first, gradually changing to a dark nut-brown, tumid, the posterior scales often larger and more prominent.

A tree of cold-temperate and subalpine levels, growing on the mountains of central and western China, and at lower altitudes in the north and in Corea. It is recognized by its tenaciously persistent cones with a remarkable change in color. It is constantly confused with P. Thunbergii and P. densiflora, neither of which grows spontaneously in China. From the former it differs in leaf-section and bud (the bud of P. sinensis is never white), from the latter in the lustre and the color variation of its cone, and from both in the frequent obliquity of its cone and in the frequent presence of trimerous leaf-fascicles.

Of the two varieties of this species, densata and yunnanensis (Shaw in Sargent, Pl. Wilson. ii. 17), the former represents the extreme oblique form of cone, the latter represents the longest dimensions of cone and leaf. The effect of environment on this species can be seen in figs. 202, 203, from a lower slope and rich soil, and fig. 204, from a high rocky ledge in the same locality.

Plate XXIII.

Fig. 201, Cone of var. densata. Fig. 202, Cone of var. yunnanensis. Fig. 203, Leaf-fascicle and magnified leaf-section of var. yunnanensis. Fig. 204, Cone and leaf-fascicle from a rocky ledge. Fig. 205, Cone, leaf-fascicle and magnified leaf-section of the typical form. Fig. 206, Seeds. Fig. 207, Conelet and its enlarged scale.

## 36. PINUS INSULARIS

- 1837 P. taeda Blanco, Fl. Filip. 767 (not Linnaeus).
- 1847 P. insularis Endlicher, Syn. Conif. 157.
- 1854 P. khasiana Griffith, Notul. Pl. Asiat. iv. 18; Icon. Pl. Asiat. tt. 367, 368.
- 1868 P. kasya Royle ex Parlatore in DC. Prodr. xvi-2, 390.

Spring-shoots uninodal, glabrous. Leaves from 12 to 24 cm. long, in fascicles of 3, rarely of 2, very slender; resin-ducts external, rarely with a medial duct. Conelets mucronate. Cones from 5 to 10 cm. long, ovate-conic, symmetrical or oblique, tenaciously persistent; apophyses lustrous, nut-brown, convex or elevated along a transverse keel, the posterior scales of some cones larger and more prominent than the anterior scales, the mucro usually deciduous.

A species of the Philippines and of northern Burmah. In both countries it is locally exploited for wood and resin. It differs from the common form of P. sinensis by its much longer leaves, and from its var. yunnanensis, which it more resembles, by its much more slender and pliant leaves. Moreover its cone, so far as I can learn, is not yellow at maturity, but brown.

Plate XXIII.

Figs. 208, 209, Three cones. Fig. 210, Leaf-fascicle and magnified leaf-section.

61

PLATE XXIII. P. MERKUSII (198-200), SINENSIS (201-207), INSULARIS (208-210)

## XI. AUSTRALES

Pits of the ray-cells small. Leaf-hypoderm biform or variable. Spring-shoots uninodal in some, multinodal in other species. Cones dehiscent at maturity.

This group combines the dehiscent cone of the Lariciones with the wood-anatomy of the serotinous Pines. Also the multinodal spring-shoot first appears here and is gradually developed among the species, absent in Nos. 37-39, sometimes present in Nos. 40-43, and prevalent in Nos. 44-47.

All the species are of the Western Hemisphere, and among them may be found the biform hypoderm of the leaf, the internal resin-duct, and the total absence of external resin-ducts, characters common in American Hard Pines. The eastern species are quite constant in their characters and present no varietal forms; the western species, on the other hand, are very variable. This difference may be due to the even level and slight climatic differences of the Atlantic states and to the remarkable diversity of altitude and climate of the western states and Mexico.

Outer walls of the leaf-endoderm thick.
Cones large, attaining 12 cm. or more in length.
Prickles of the cone inconspicuous.

| | |
|---|---|
| Bark-formation late | 37. pseudostrobus |
| Bark-formation early | 38. Montezumae |
| Prickle of the cone conspicuous | 39. ponderosa |
| Cones small, 7 cm. or less in length | 40. teocote |

Outer walls of the leaf-endoderm thin.
Spring-shoots mostly uninodal.
Prickle of the cone slender, sometimes deciduous.

| | |
|---|---|
| Cones mostly oblique | 41. Lawsonii |
| Cones symmetrical | 42. occidentalis |
| Prickles of the cone stout and persistent | 43. palustris |

Spring-shoots multinodal.

| | |
|---|---|
| Resin-ducts internal | 44. caribaea. |

| | |
|---|---|
| Resin-ducts mostly medial. | |
| Prickle of the cone stout | 45. taeda |
| Prickle of the cone slender. | |
| Bark-formation late | 46. glabra |
| Bark-formation early | 47. echinata |

## 37. PINUS PSEUDOSTROBUS

- 1839 P. pseudostrobus Lindley in Bot. Reg. xxv. Misc. 63.
- 1839 P. apulcensis Lindley in Bot. Reg. xxv. Misc. 63.
- 1842 P. tenuifolia Bentham, Pl. Hartw. 92.
- 1846 P. orizabae Gordon in Jour. Hort. Soc. Lond. i. 237, f.

Spring-shoots uninodal, conspicuously pruinose. Bark-formation late, the cortex of young trees smooth. Leaves in fascicles of 5, sometimes of 6, from 15 to 30 cm. long, drooping; resin-ducts medial, hypoderm variable in amount, often in very large masses, the outer walls of the endoderm thick. Conelets mucronate. Cones from 7 to 14 cm. long, ovate or ovate-conic, symmetrical or oblique, deciduous and often leaving a few basal scales on the trees; apophyses rufous or fulvous brown, flat, elevated or, in one variety, prolonged in various degrees, the prolongations nearly uniform or much more prominent on the posterior face of the cone, the mucro usually deciduous.

A species of the subtropical and warm-temperate altitudes of Mexico and Central America. Its range includes both eastern and western slopes of the northern plateau. Its northern limit is in Nuevo Leon, and it probably reaches in Nicaragua the southern limit of pines in the Western Hemisphere. It is distinguished from all its associates by the smooth gray trunk of the young trees, by their long internodes, and by their drooping gray-green foliage.

64

Some cones of this species develop protuberances of all degrees of prominence up to the curious cone collected in Oaxaca by Nelson (var. apulcensis, Shaw, Pines Mex. t. 12, fig. 8). There is also a remarkable difference in the amount of leaf-hypoderm. On many trees of the western part of the range this tissue forms septa across the green mesophyll. Such partitions are sometimes met in other species, P. Pringlei or P. canariensis, where the hypoderm is abundant. But in P. pseudostrobus they appear in some leaves of weak, as well as of strong hypoderm (var. tenuifolia, Shaw, Pines Mex. t. 13, ff. 2, 4, 5, 7, 8).

Plate XXIV.

Fig. 211, Cone. Fig. 212, Two cones of var. tenuifolia. Figs. 213, 214, Two cones of var. apulcensis. Fig. 215, Magnified section of 3 leaves of var. tenuifolia. Fig. 216, Magnified section of 2 leaves of the species. Fig. 217, Bud destined to produce staminate flowers. Fig. 218, Ten-year old branch showing smooth cortex. Fig. 219, Young and mature trees in open growth.

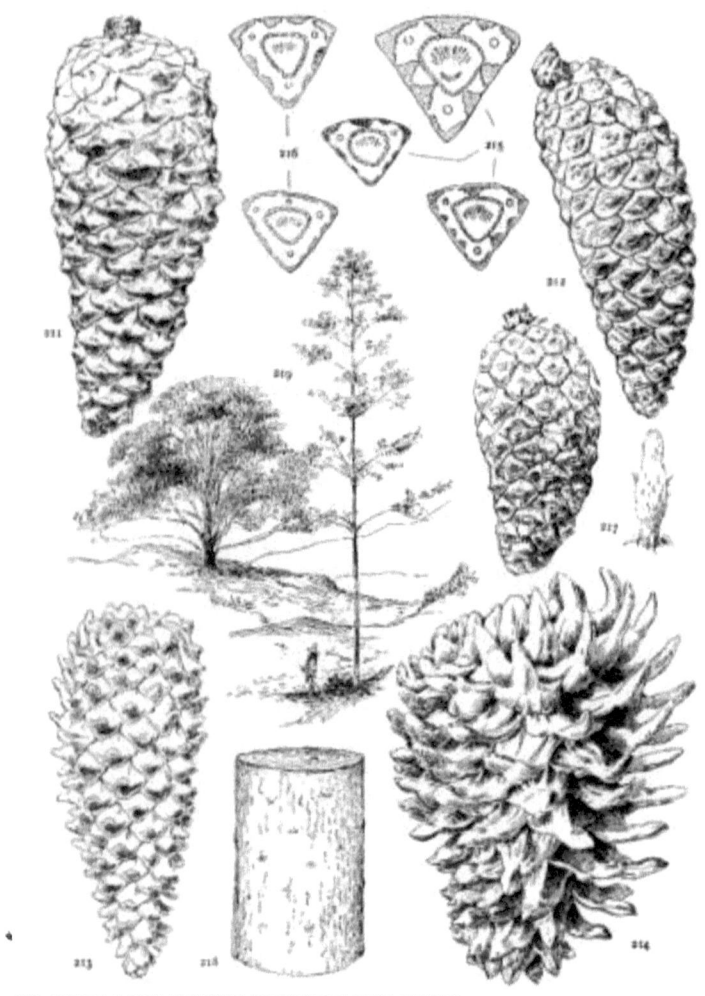

PLATE XXIV. PINUS PSEUDOSTROBUS

## 38. PINUS MONTEZUMAE

- 1817 P. occidentalis H. B. & K. Nov. Gen. ii. 4 (not Swartz).
- 1832 P. Montezumae Lambert, Gen. Pin. ed. 8vo, i. 39, t. 22.
- 1839 P. Devoniana Lindley in Bot. Reg. xxv. Misc. 62.

- 1839 P. Hartwegii Lindley in Bot. Reg. xxv. Misc. 62.
- 1839 P. Russelliana Lindley in Bot. Reg. xxv. Misc. 63.
- 1839 P. macrophylla Lindley in Bot. Reg. xxv. Misc. 63.
- 1840 P. filifolia Lindley in Bot. Reg. xxvi. Misc. 61.
- 1841 P. Sinclairii Hooker & Arnott, Bot. Beechy Voy. 392, t. 93 (as to cone).
- 1841 P. radiata Hooker & Arnott, Bot. Beechy Voy. 443 (as to leaves).
- 1847 P. Grenvilleae Gordon in Jour. Hort. Soc. Lond. ii. 77, f.
- 1847 P. Gordoniana Hartweg in Jour. Hort. Soc. Lond. ii. 79, f.
- 1847 P. Wincesteriana Gordon in Jour. Hort. Soc. Lond. ii. 158, f.
- 1847 P. rudis Endlicher, Syn. Conif. 151.
- 1847 P. Ehrenbergii Endlicher, Syn. Conif. 151.
- 1858 P. Lindleyana Gordon, Pinet. 229.
- 1891 P. Donnell-Smithii Masters in Bot. Gaz. xvi. 199.

Spring-shoots uninodal, slightly or not at all pruinose. Bark-formation early, the branches becoming dark and rough. Leaves prevalently in fascicles of 5, but varying from 3 to 8, extremely variable in length, attaining 45 cm. at subtropical levels; resin-ducts medial, hypoderm sometimes uniform, more commonly multiform, the outer walls of the endoderm thick. Conelet mucronate, the prickle often reflexed. Cones of many sizes, attaining in warm localities 30 cm. in length, ovate-conic or long-conic, symmetrical, often curved, deciduous and often leaving a few scales on the tree; apophyses dull, rarely lustrous, nut-brown, or of various shades of fuscous brown to nearly black, flat, tumid, pyramidal or sometimes slightly protuberant, the prickle rarely persistent.

This species ranges from the mountains of northern Durango to the volcanoes of Guatemala, or possibly farther south. It is found at all altitudes where Pines can grow except on the tropical levels of Guatemala. Its more hardy forms have been successfully grown in the milder parts of Great Britain and northern Italy. It is felled for lumber in many parts of Mexico.

This sturdy Pine and its numberless variations present the most remarkable example of adaptation in the genus. The variations are mostly those associated with changes of environment—dimensions of cone and leaf and the number of leaves in the fascicle. These are so accurately correlated with altitude and exposure, and are so imperceptibly graded, that no specific segregations among them have yet been successfully established.

The type-specimen figured by Lambert does not show the longest cone and leaf of this species. They are better represented by specimens which have been named P. filifolia. Such dimensions prevail in subtropical localities. At temperate altitudes these dimensions are much reduced, but here are found a longer form of cone and leaf (var. Lindleyi, Loudon) and a shorter form (var. rudis, 66 Shaw). At still higher altitudes and up to the timber-limit the var. Hartwegii, Engelmann, with short leaves and a small nearly black cone is found. Among these varieties there is no such sharp distinction as these definitions imply. All dimensions of fruit and foliage and the various brown and black shades of the cone blend into each other through endless intergradations. A monograph of this species, by one who could devote some years to it on the superb volcanoes and in the delightful climates where this tree abounds, would be a valuable contribution to science.

Plate XXV. (Cones and leaves much reduced.)

Fig. 220, Cone and leaves of Lambert's plate. Figs. 221, 222, Longer cones and leaves of the species. Fig. 223, Cone and leaves of var. Lindleyi. Fig. 224, Cones and leaves of var. rudis. Fig. 225, Cone and leaves of var. Hartwegii. Fig. 226, Magnified leaf-sections. Figs. 227, 228, Two forms of the dermal tissues of the leaf, magnified. Fig. 229, Habit of the tree.

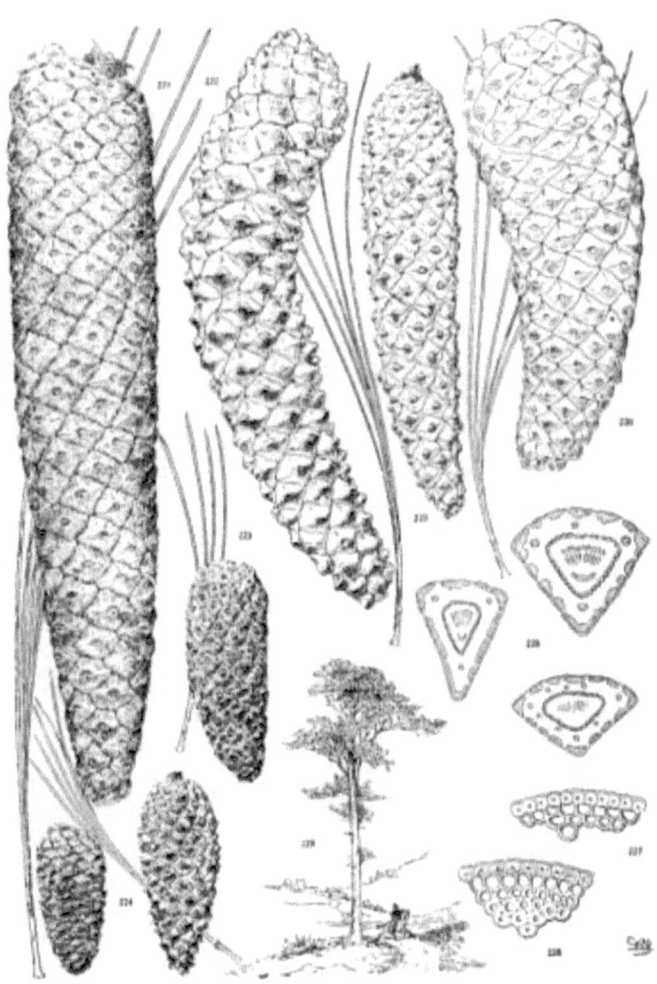

PLATE XXV. PINUS MONTEZUMAE

## 39. PINUS PONDEROSA

- 1836 P. ponderosa Douglas ex Lawson's Agric. Man. 354.
- 1847 P. Benthamiana Hartweg in Jour. Hort. Soc. Lond. ii. 189.

- 1848 P. brachyptera Engelmann in Wislizenus, Tour Mex. 89.
- 1848 P. macrophylla Engelmann in Wislizenus, Tour Mex. 103 (not Lindley).
- 1853 P. Jeffreyi Balfour in Bot. Exp. Oregon, 2, f.
- 1854 P. Engelmanni Carrière in Rev. Hort. 227.
- 1855 P. Beardsleyi Murray in Edinb. Phil. Jour. ser. 2, i. 286, t. 6.
- 1855 P. Craigana Murray in Edinb. Phil. Jour. ser. 2, i. 288, t. 7.
- 1858 P. Parryana Gordon, Pinet. 202 (not Engelmann).
- 1859 P. deflexa Torrey in Emory's Rep. ii. 1, 209, t. 56.
- 1878 P. arizonica Engelmann in Wheeler's Rep. vi. 260.
- 1889 P. latifolia Sargent in Gar. & For. ii. 496, f. 135.
- 1894 P. apacheca Lemmon in Erythea, ii. 103, t. 3.
- 1897 P. Mayriana Sudworth in Bull. 14, U. S. Dept. Agric. 21.
- 1897 P. scopulorum Lemmon in Gar. & For. x. 183.
- 1900 P. peninsularis Lemmon, W. Am. Conebear. 114.

Spring-shoots uninodal, sometimes pruinose. Bark-formation early. Leaves prevalently in fascicles of 3, but varying from 2 to 5 or more, from 12 to 36 cm. long; resin-ducts medial, hypoderm uniform or multiform, outer walls of the endoderm thick. Conelet mucronate, the mucro often reflexed. Cones from 8 to 20 cm. long, ovate-conic, symmetrical, deciduous and usually leaving a few basal scales on the tree; apophyses tawny yellow to fuscous brown, lustrous, elevated along a transverse keel, sometimes protuberant and reflexed, the umbo salient and forming the base of a pungent, persistent prickle.

This species ranges from southern British Columbia over the mountains between the Pacific and the eastern foot-hills of the Rocky Mountains, including the Black Hills of South Dakota, to the northeastern Sierras of Mexico, to northern Jalisco and Lower California, forming, in many localities, large forests and furnishing the best Hard Pine timber of the western United States. It attains its best growth on the Sierras of California and is, next to P. Lambertiana, the tallest of the Pines.

Like P. Montezumae, and under like influences, it shows much dimensional variation, and the leaf-fascicles are heteromerous, with the larger number in the southern part of its range. Many authors consider the variety Jeffreyi Vasey to be a distinct species; but here, it seems to me, too much importance is attached to the pruinose branchlet, clearly a provision against transpiration and associated rather with a dry environment than with a species. Most observers discover many intermediate forms between this variety and the species. The var. scopulorum Engelm. is the Rocky Mountain form with leaves in 2's and 3's and with small cones passing into P. arizonica, Engelm., a more southern form with small cones and leaves in fascicles of 3 to 5. The var. macrophylla (Shaw, Pines Mex. 24), in addition to its long and stout leaves, bears a cone with protuberant apophyses, somewhat comparable to the intermediate forms of P. pseudostrobus var. apulcensis 68 Shaw (l. c.). Fascicles of 6 and 7 leaves are sometimes found, and specimens that I have collected in Sandia, Durango (issued by Pringle, through a misunderstanding, under the name P. Roseana, ined.) show such fascicles on the fertile branches.

Plate XXVI.

Fig. 230, Cone and seed of var. Jeffreyi. Fig. 231, Cone of var. macrophylla. Fig. 232, Cone of var. scopulorum. Fig. 233, Magnified leaf-section and cells of leaf-endoderm. Fig. 234, Magnified dermal tissues of the leaf, showing uniform and multiform hypoderm.

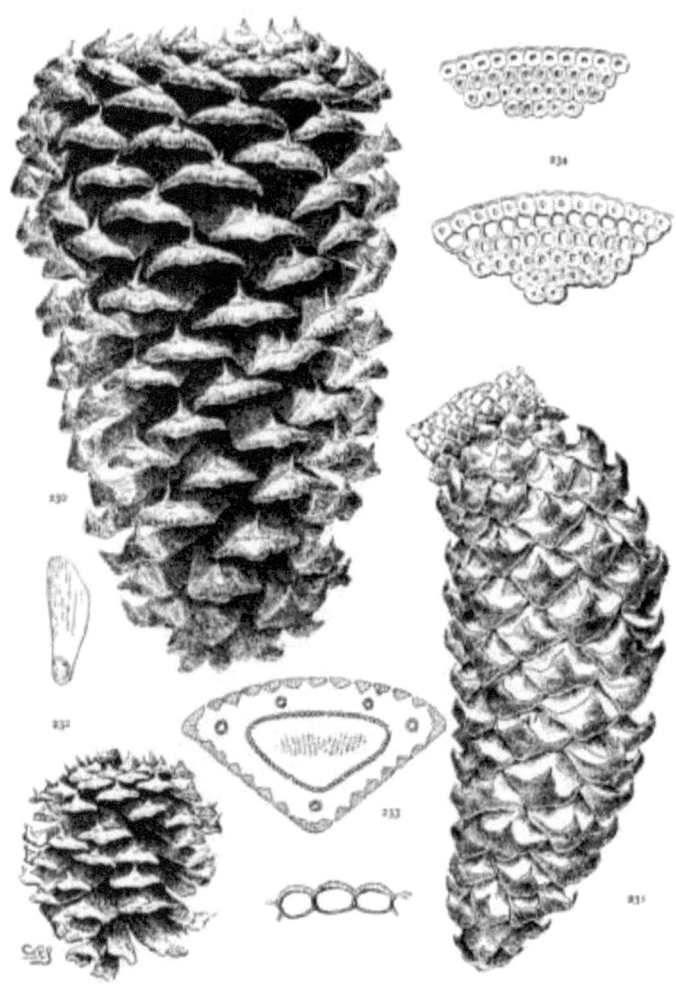

PLATE XXVI. PINUS PONDEROSA

## 40. PINUS TEOCOTE

- 1830 P. teocote Schlechtendal & Chamisso in Linnaea, v. 76.

Spring-shoots uninodal, or sometimes multinodal. Leaves prevalently in fascicles of 3, but varying from 3 to 5, from 10 to 20 cm. long; resin-ducts medial, sometimes with an internal duct, hypoderm biform, endoderm with thick outer walls. Conelets mucronate. Cones usually very small, from 4 to 6 cm. long, but with a larger varietal form, ovate to long-conic, symmetrical; apophyses nut-brown, flat or tumid, the mucro usually deciduous.

This species grows at temperate altitudes from Chiapas to Nuevo Leon, associated with temperate Mexican species such as P. patula, P. leiophylla and others, and is easily recognized by its small cone. The variety with a larger cone (var. macrocarpa, Shaw, Pines Mex. t. 10) I have found growing in mixed groves of P. teocote and P. leiophylla. It resembles the latter in cone and leaf, but lacks the peculiar character that distinguishes P. leiophylla from all other Mexican species—the triennial cone. Some of the specimens of Hartweg No. 441 belong here, as well as Pringle's specimens, Nos. 10013, 10018, distributed as P. eslavae, ined.

Plate XXVII.

Fig. 235, Two cones of the species and the larger cone of the variety. Fig. 236, Leaf-fascicle and magnified sections of two leaves. Fig. 237 a, Dermal tissues of the leaf magnified; b, magnified cells of the leaf-endoderm. Fig. 238, Habit of the tree.

### 41. PINUS LAWSONII

- 1862 P. Lawsonii Roezl ex Gordon, Pinet. Suppl. 64.
- 1905 P. Altamirani Shaw in Sargent, Trees & Shrubs, i. 209, t. 99.

Spring-shoots conspicuously pruinose, uninodal or not infrequently multinodal. Leaves in fascicles of 3, 4 or 5, not exceeding 24 cm. in length; resin-ducts internal, often with one or two medial ducts, hypoderm biform, endoderm usually with thin outer walls. Conelets mucronate. Cones from 5 to 7 cm. long on pliant peduncles, ovate or ovate-conic, oblique or sometimes symmetrical, deciduous, or persistent with a weak hold on the branch; apophyses nut-brown, flat or tumid, often protuberant on the posterior face of the cone, the umbo usually large and salient, forming a rounded button-like projection, on which the mucro is wanting.

A subtropical species of central and western Mexico, growing alone or associated with P. oocarpa, P. Pringlei and the subtropical forms of P. Montezumae and P. pseudostrobus. It is recognized among its associate species by its conspicuously glaucous foliage. The cone is very variable on trees of the same grove, both in size and in the protuberance of its apophyses. Gordon's specimen in the Kew herbarium consists of a single detached cone and a few leaves. The leaves differ from all that I have examined in showing thick-walled endoderm cells, but the cone corresponds with many of my own collection.

Plate XXVII.

Fig. 239, Three cones. Fig. 240, Leaf-fascicle and magnified leaf-section. Fig. 241, Magnified cells of the leaf-endoderm.

69

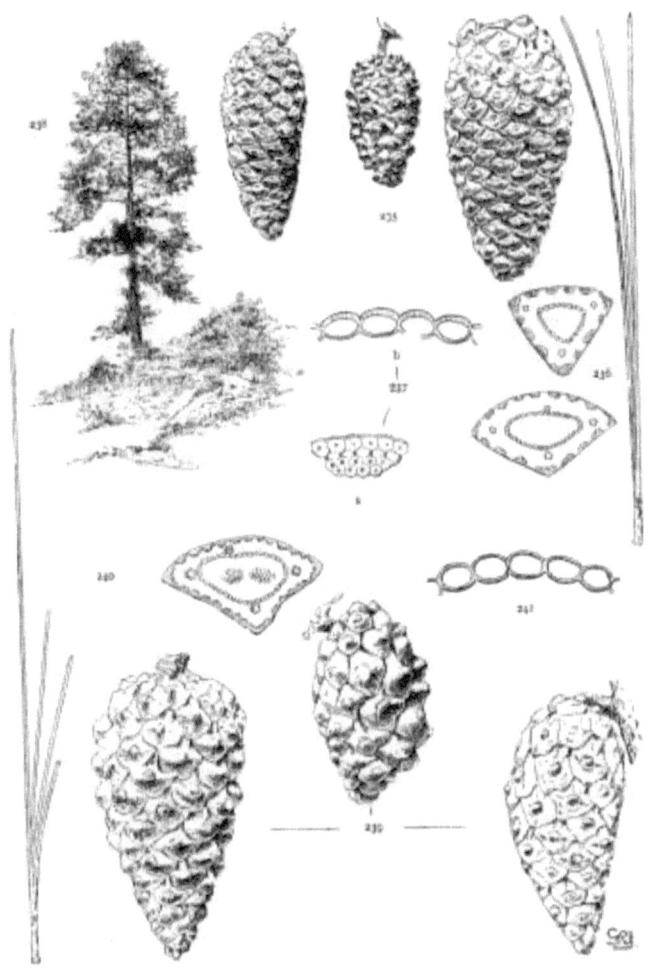

PLATE XXVII. P. TEOCOTE (235-238), LAWSONII (239-241)

## 42. PINUS OCCIDENTALIS

- 1788 P. occidentalis Swartz, Nov. Gen. & Sp. Pl. 103.

- 1862 P. cubensis Grisebach in Mem. Am. Acad. ser. 2, viii. 530.
- 1880 P. Wrightii Engelmann in Trans. Acad. Sci. St. Louis, iv. 185.

Spring-shoots uninodal, pruinose. Leaves in fascicles of 2 to 5, from 15 to 22 cm. long; resin-ducts internal, hypoderm biform, endoderm with thin outer walls. Conelets erect, aristate. Cones from 5 to 8 cm. long, reflexed, ovate, symmetrical, deciduous; apophyses nut-brown, lustrous, flat or tumid, the umbo often thin and, together with the slender prickle, bent sharply downward.

This species is confined to San Domingo, Hayti and eastern Cuba. Its erect conelet and reflexed cone distinguish it from P. caribaea, which has both its conelet and cone reflexed. Moreover the conelet is usually, perhaps always, subterminal in P. occidentalis.

Plate XXVIII.

Fig. 247, Cone. Fig. 248, Conelet and enlarged aristate scales. Fig. 249, Magnified sections of two leaves and more magnified dermal tissues.

### 43. PINUS PALUSTRIS

- 1768 P. palustris Miller, Gard. Dict. ed. 8.
- 1810 P. australis Michaux f. Hist. Arbr. Am. i. 64, t. 6.

Spring-shoots uninodal, rarely multinodal. Buds peculiarly large, white, and conspicuously fringed with the long free cilia of the bud-scales. Leaves in fascicles of 3, from 20 to 45 cm. long, rigid; resin-ducts internal, hypoderm biform, endoderm with thin outer walls. Conelets short-mucronate. Cones from 15 to 20 cm. long, narrow, tapering from a rounded base to a blunt point, symmetrical, deciduous and usually leaving a few scales on the tree; apophyses dull nut-brown, elevated along a transverse keel, the umbo salient and forming the broad base of a small persistent prickle.

Its thin sap-wood, its very strong heavy wood of large dimensions with abundant resin of excellent quality make this the most valuable species of the genus. It ranges over the sandy plain that borders the Atlantic and the Gulf of Mexico, from southeastern Virginia to eastern Texas. The northern limit is approximately the centre of the Southern and Gulf States, with a northern extension in Alabama to the base of the Appalachian Mountains and to northwestern Louisiana. Its southern limit lies near the centre of the Florida peninsula.

Among its associates this species is recognized by its large white fringed bud and its elongated cone. Its leaves attain, on vigorous trees, the maximum length among Pines, but on most trees the leaves do not differ in length from the longer forms of those of P. caribaea or P. taeda. A peculiarity, which it shares with P. caribaea, is the deciduous scaly bark of mature trees, constantly falling away in thin irregular scales.

Plate XXVIII.

Figs. 242, 243, Cones and seed. Fig. 244, Bud. Fig. 245, Magnified leaf-section. Fig. 246, Magnified cells of the leaf-endoderm. The dermal tissues of fig. 249 also apply to this species.

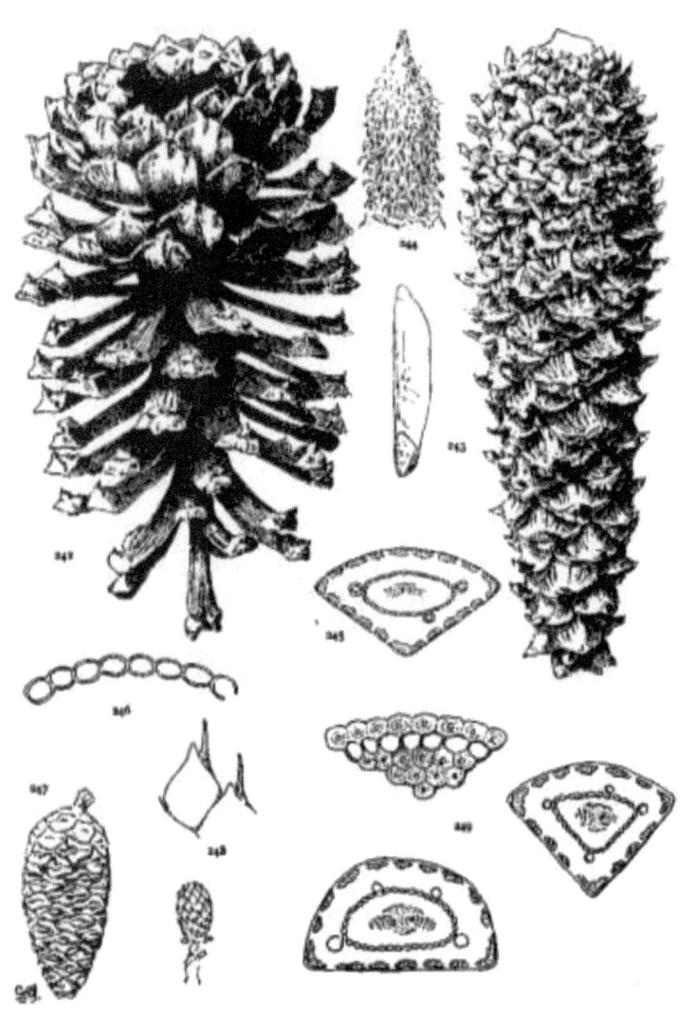

PLATE XXVIII. P. PALUSTRIS (242-246), OCCIDENTALIS (247-249)

## 44. PINUS CARIBAEA

- 1851 P. caribaea Morelet in Rev. Hort. Côte d'Or, i. 105.
- 1864 P. bahamensis Grisebach, Fl. Brit. W. Ind. 503.

- 1880 P. Elliottii Engelmann in Trans. Acad. St. Louis, iv. 186, tt. 1-3.
- 1884 P. cubensis Sargent in Rep. 10th. Cens. U. S. ix. 202 (not Grisebach).
- 1893 P. heterophylla Sudworth in Bull. Torrey Bot. Club, xx. 45.
- 1903 P. recurvata Rowley in Bull. Torrey Bot. Club, xxx. 107.

Spring-shoots multinodal, more or less pruinose. Buds pale chestnut-brown. Leaves in fascicles of 2 and 3, or more in its southern range, from 12 to 25 cm. long; resin-ducts internal, hypoderm biform, endoderm with thin outer walls. Conelets reflexed on long peduncles, mucronate. Cones 72 from 5 to 15 cm. long, ovate or oblong-ovate, symmetrical, deciduous and leaving often a few basal scales on the branch; apophyses lustrous, rufous-brown, tumid, the umbo somewhat salient and minutely mucronate.

The northern limit of the range of P. caribaea extends from the coast of southeastern S. Carolina through southeastern Georgia and southern Alabama to southeastern Louisiana. It is associated with P. palustris, taeda, serotina, echinata and glabra in this part of its range. It continues through Florida, where it encounters P. clausa. On the Bahamas it is the only Pine. On the Isle of Pines it finds in P. tropicalis another associate. It also grows in Honduras and Guatemala. The wood and resin of this species are of such excellent quality that no commercial distinction is made between P. caribaea and P. palustris.

Plate XXIX.

Fig. 250, Cone from the Isle of Pines. Fig. 251, Small form of cone. Fig. 252, Large form of cone and binate leaf-fascicle. Fig. 253, Conelet. Fig. 254, Magnified sections of leaves from binate and ternate fascicles. Fig. 255, Habit of the tree, contrasted with a tree of P. palustris in the middle-distance.

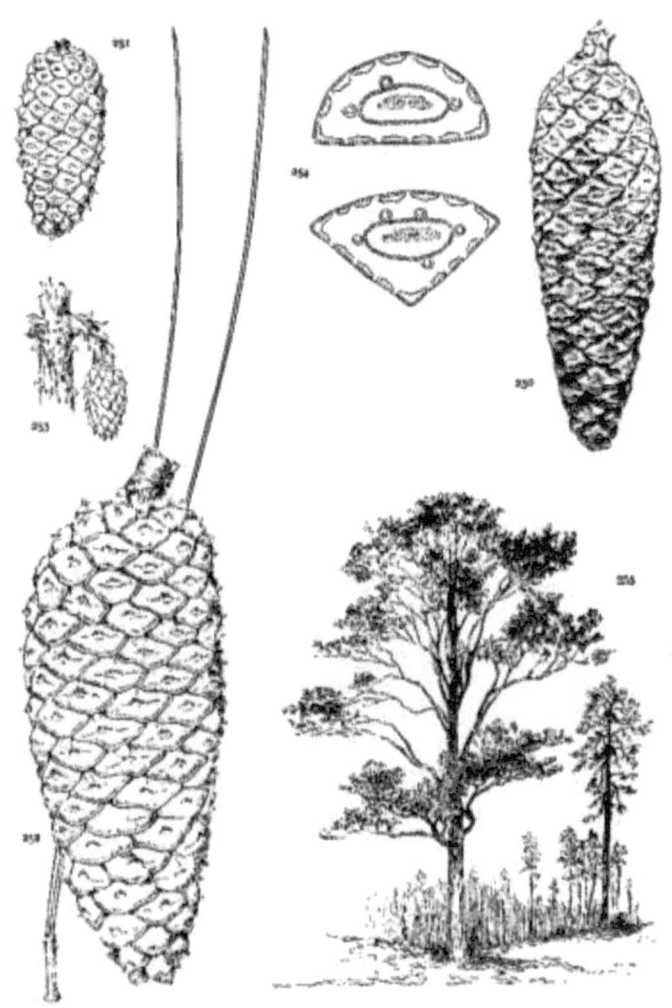

PLATE XXIX. PINUS CARIBAEA

## 45. PINUS TAEDA

- 1753 P. taeda Linnaeus, Sp. Pl. 1000.
- 1788 P. lutea Walter Fl. Carol. 237.

- 1903 P. heterophylla Small, Fl. Southeast. U. S. 28 (not Sudworth).

Spring-shoots multinodal. Leaves in fascicles of 3, from 12 to 25 cm. long; resin-ducts medial, sometimes with an internal duct, hypoderm biform, endoderm with thin outer walls. Conelets erect, their scales prolonged into a sharp point. Cones from 6 to 10 cm. long, ovate-conic, symmetrical; apophyses dull pale nut-brown, rarely lustrous, elevated along a transverse keel, the whole umbo forming a stout triangular spine with slightly concave sides.

The species ranges from southern New Jersey to southern Arkansas, Oklahoma, eastern Texas and southwestern Tennessee, but does not occur in the lower half of the Florida peninsula. It is an important timber-tree, manufactured into all descriptions of scantlings, boarding and finish, but the wood is of various qualities. It may be recognized by the spine of its cone in both years of growth. Excepting the formidable armature of the cone of P. pungens, the spines are the strongest and most persistent of all the species of eastern North America.

Plate XXX.

Fig. 264, Cone. Fig. 265, Leaf-fascicle. Fig. 266, Magnified leaf-section. Fig. 267. Magnified scales of the conelet.

## 46. PINUS GLABRA

- 1788 P. glabra Walter, Fl. Carol. 237.

Spring-shoots multinodal. Bark-formation late, the upper trunks of mature trees smooth. Leaves in fascicles of 2, from 9 to 12 cm. long; resin-ducts medial, hypoderm weak, sometimes of a single row, biform when of two rows, endoderm with thin outer walls. Conelets reflexed, mucronate. Cones from 4 to 7 cm. long, reflexed, ovate, symmetrical, deciduous on some trees, persistent on others; apophyses pale dull nut-brown, thin or slightly thickened, the prickle usually deciduous.

A tree that sometimes attains important dimensions, growing singly or in small groves from the neighborhood of Charleston, S. C., to eastern Louisiana and central Mississippi, most abundant in a strip of territory on either side of the northern boundary of Florida. Among the Pines of the southeastern United States it is the only species with late bark-formation, and is therefore easily identified.

Plate XXX.

Fig. 256, Cone. Fig. 257, Enlarged scale of the conelet. Fig. 258, Leaf-fascicle and magnified leaf-section. Fig. 259, Dermal tissues of the leaf magnified, with a double row of hypoderm cells.

## 47. PINUS ECHINATA

- 1768 P. echinata Miller, Gard. Dict. ed. 8.
- 1788 P. squarrosa Walter, Fl. Carol. 237.
- 1803 P. mitis Michaux, Fl. Bor. Am. ii. 204.
- 1803 P. variabilis Lambert, Gen. Pin. i. 22, t. 15.
- 1854 P. Royleana Jamieson in Jour. Hort. Soc. Lond. ix. 52, f.

Spring-shoots multinodal, somewhat pruinose. Bark forming early, rough on the upper trunk. Leaves in fascicles of 2 and 3, from 7 to 12 cm. long; resin-ducts medial, with an occasional internal duct, hypoderm weak, biform when of two rows of cells, endoderm with thin outer walls. Conelets mucronate. Cones from 4 to 6 cm. long, ovate-conic, symmetrical, often persistent; apophyses dull pale nut-brown, thin or somewhat thickened along a transverse keel, the umbo salient, the mucro more or less persistent.

This species ranges from southeastern New York to northern Florida, to West Virginia and eastern Tennessee, and through the Gulf States to eastern Louisiana, eastern Texas, southern Missouri and southwestern Illinois. It is extensively manufactured into material of all kinds that enters into the construction of buildings. It differs from P. virginiana in its longer leaves, brittle branches, and much greater height, from P. glabra in its rough upper trunk, and from both by the frequent presence of trimerous leaf-fascicles.

Of the six or seven pines of the southeastern United States, this species covers a larger area and ascends the slopes of the Alleghany Mountains far enough to meet the northern species, P. virginiana, P. rigida, and P. strobus. Unlike the western members of this group, P. echinata and its associates are not variable. Their characters are singularly constant, as their limited synonymy and total lack of varietal names attest.

Plate XXX.

Fig. 260, Cone. Fig. 261, Leaf-fascicle and magnified leaf-section from a ternate fascicle. Fig. 262, Magnified leaf-section from a binate fascicle. Fig. 263, Multinodal branchlet bearing lateral and subterminal conelets and a ripe cone. Figs. 257, showing mucronate scales of the conelet, and 259, showing dermal tissues of the leaf, are applicable also to this species.

75

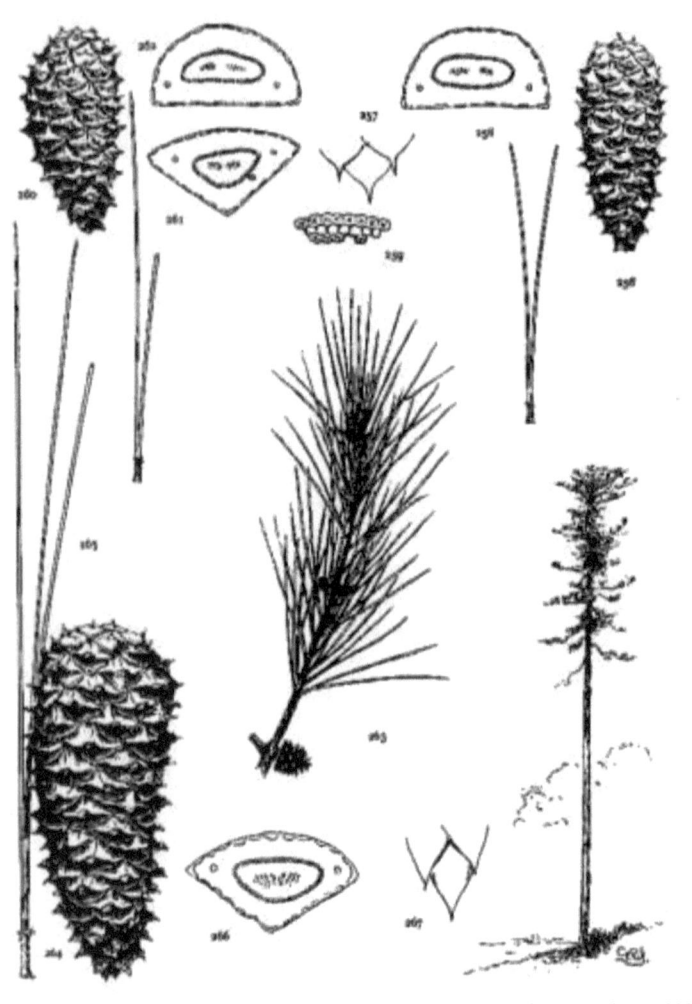

PLATE XXX. P. GLABRA (256-259), ECHINATA (260-263), TAEDA (264-267)

## XII. INSIGNES

Pits of the ray-cells small. Cones tenaciously persistent, serotinous in various degrees. Conelets mucronate or spinose.

Spring-shoots uninodal.
Resin-ducts mostly internal            48. Pringlei
Resin-ducts mostly septal               49. oocarpa
Spring-shoots multinodal.
Cones symmetrical.
Leaf-hypoderm not biform.
Bark-formation late                      50. halepensis
Bark-formation early                    51. pinaster
Leaf-hypoderm biform.
Cones with slender spines.
Leaves binate.
Cones dehiscent at maturity         52. virginiana
Cones serotinous                      53. clausa
Leaves ternate.
Cones dehiscent at maturity         54. rigida
Cones serotinous                      55. serotina
Cones with stout spines              56. pungens
Cones oblique or unsymmetrical.
Cones and leaves very short, not exceeding 6 cm.
Cones curved or warped             57. Banksiana
Cones straight                         58. contorta
Cones and leaves much longer, more than 7 cm.
Posterior cone-scales gradually larger than anterior scales.
Bark-formation late                      59. Greggii
Bark-formation early                    60. patula
Posterior cone-scales abruptly larger than anterior scales.
Cones with very stout spines        61. muricata
Cones with minute or deciduous prickles.

| Bark-formation late | 62. attenuata |
| Bark-formation early | 63. radiata |

## 48. PINUS PRINGLEI

1905 P. Pringlei Shaw in Sargent, Trees & Shrubs, i. 211, t. 100.

Spring-shoots uninodal, sometimes pruinose. Leaves ternate, from 15 to 25 cm. long; resin-ducts internal or with an occasional septal duct, hypoderm biform, in thick masses, often projecting far into the green tissue and sometimes touching the endoderm. Conelets mucronate. Cones from 5 to 10 cm. long, reflexed on a rigid peduncle, subsymmetrical or more or less oblique, tenaciously persistent, often serotinous; apophyses sublustrous tawny yellow or fulvous brown, convex, the posterior scales often more prominently developed, the mucro usually wanting; seed with a perceptibly thickened wing-blade.

A tree with long erect bright green foliage, confined, so far as known, to the subtropical altitudes of western Mexico. As it grows in Uruapan, Michoacan, there are two forms of the cone, large and small, both with the same long rigid leaf.

Plate XXXI.

Figs. 268, 269. Three cones and seed. Fig. 270, Leaf-fascicle and magnified leaf-section.

78

## 49. PINUS OOCARPA

- 1838 P. oocarpa Schiede in Linnaea, xii. 491.
- 1842 P. oocarpoides Lindley ex Loudon, Encycl. 1118.

Spring-shoots uninodal, pruinose. Leaves in fascicles of 3, 4 or 5, from 15 to 30 cm. long, erect; resin-ducts mostly septal, sometimes internal, hypoderm biform or multiform. Conelets on very long peduncles, mucronate. Cones from 4 to 10 cm. long, long-pedunculate, broad-ovate to ovate-conic, symmetrical or sometimes oblique, persistent, more or less serotinous; apophysis gray-yellow or greenish yellow of high lustre, flat or variously convex, delicately and radially carinate, the umbo often salient, the prickle usually broken away; seed-wing appreciably thickened at the base of the blade.

A subtropical species, ranging from Guatemala to the northern border of Sinaloa in northern Mexico; remarkable for the length of the peduncle of the cone and for the prevalence of septal resin-ducts in the leaf.

Plate XXXI.

Fig. 271, Three cones and seed. Fig. 272, Leaf-fascicle and magnified leaf-section. Fig. 273, Cone from northern part of the range. Fig. 274, Leaf-fascicle and magnified leaf-section from near the northern limit.

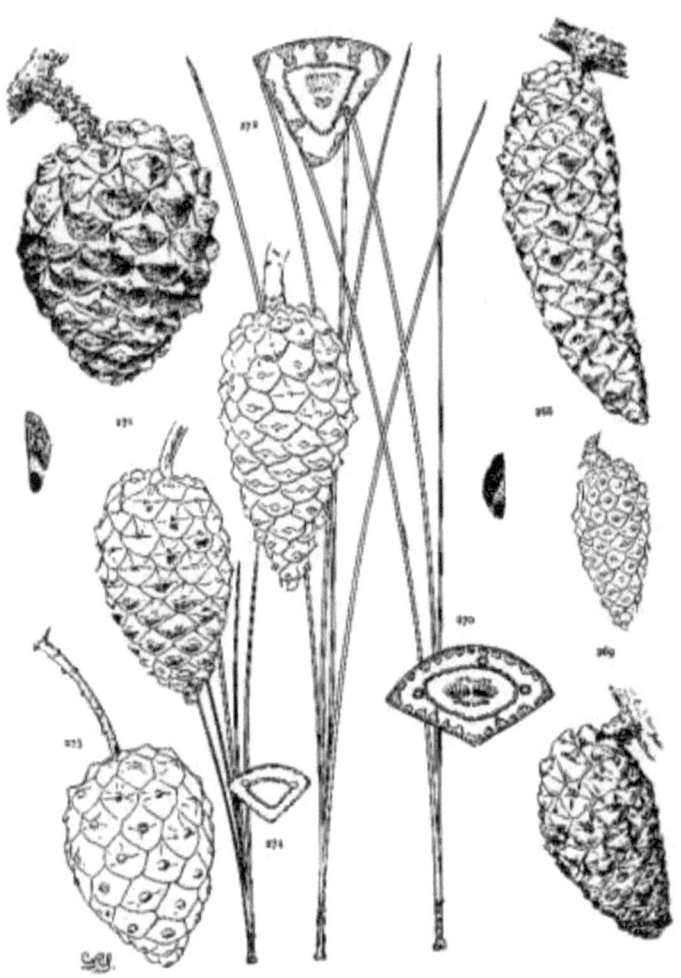

PLATE XXXI. P. PRINGLEI (268-270), OOCARPA (271-274)

## 50. PINUS HALEPENSIS

- 1762 P. sylvestris Gouan, Hort. Reg. Monspel. 494 (not Linnaeus).
- 1768 P. halepensis Miller, Gard. Dict. ed. 8.

- 1803 P. maritima Lambert, Gen. Pin. i. 13, t. 10.
- 1812 P. resinosa Loiseleur, Nouv. Duham. v. 237, t. 77 (not Aiton).
- 1815 P. brutia Tenore, Cat. Hort. Neap. Appx. 1, 75.
- 1826 P. arabica Sieber ex Sprengel, Syst. Veg. iii. 886.
- 1833 P. pyrenaica David in Ann. Soc. Hort. Paris, 186 (not Lapeyrouse).
- 1834 P. hispanica Cook, Sketches in Spain, ii. 337.
- 1838 P. pityusa Steven in Bull. Soc. Nat. Mosc. xi. 49.
- 1841 P. carica Don in Ann. Mag. Nat. Hist. vii. 459.
- 1847 P. persica Strangways ex Endlicher, Syn. Conif. 157.
- 1855 P. abasica Carrière, Trait. Conif. 352.
- 1855 P. Loiseleuriana Carrière, Trait. Conif. 382.
- 1856 P. Parolinii Visiani in Mem. Ist. Venet. vi. 243, t. 1.
- 1902 P. eldarica Medwejew in Act. Hort. Tiflis. vi-2, 21, f.

Spring-shoots often multinodal. Bark-formation late, the branches ashen gray and smooth for several years. Leaves binate, from 6 to 15 cm. long; resin-ducts external, hypoderm uniform. Conelets obscurely mucronate near the apex. Cones from 8 to 12 cm. long, ovate-conic, symmetrical or subsymmetrical, persistent, often serotinous; apophyses red with a lighter or deeper brownish shade, lustrous, flat, convex or low-pyramidal, radially carinate, the umbo often ashen gray and unarmed.

A tree ranging from Portugal to Afghanistan, and from Algeria to Dalmatia and to northern Italy and Southern France. It is a vigorous species in its own home, growing readily in poor soils, but not successful in colder climates. The wood is resinous and valuable for fuel. The turpentine industry, once associated with this species, has gradually been abandoned for the more copious product of P. pinaster.

It is recognized by its lustrous red cones and by the ashen gray cortex of its branches and upper trunk. Tenore's P. brutia (pyrenaica of some authors) is founded on a difference in the length of the leaf and on an erect cone with a shorter peduncle. To recognize species on such distinctions would not be consistent with the purpose and spirit of this discussion.

Plate XXXII.

Fig. 279, Two cones. Fig. 280, Cone. Fig. 281, Lateral conelet. Fig. 282, Magnified leaf-section. Fig. 283, Dermal tissues of the leaf magnified.

## 51. PINUS PINASTER

- 1768 P. sylvestris Miller, Gard. Dict. ed. 8 (not Linnaeus).
- 1789 P. pinaster Aiton, Hort. Kew. iii. 367.
- 1798 P. laricio Savi, Fl. Pisa. ii. 353 (not Poiret).
- 1804 P. maritima Poiret in Lamarck, Encycl. Méth. v. 337 (not Lambert).
- 1826 P. escarena Risso, Hist. Nat. ii. 340.
- 1835 P. Lemoniana Bentham in Trans. Hort. Soc. Lond. ser. 2, i. 512, t.
- 1845 P. Hamiltonii Tenore, Cat. Ort. Nap. 90.

Spring-shoots sometimes multinodal. Bark-formation early. Leaves binate, from 10 to 20 cm. long, stout and rigid; resin-ducts medial, hypoderm multiform, the inner cells gradually larger, remarkably large in the angles of the leaf. Conelets minutely mucronate. Cones from 9 to 18 cm. long, nearly sessile, ovate-conic, symmetrical or subsymmetrical, persistent, sometimes serotinous; apophyses lustrous nut-brown or rufous brown, conspicuously pyramidal, the umbo salient and pungent.

A maritime tree corresponding nearly, in its range, with the preceding species, but more hardy in cooler climates. It grows from Portugal to Greece, and from Algeria to Dalmatia, but its area has been much extended by cultivation. Under favorable conditions it attains large dimensions, but its exploitation for resin and turpentine tends to diminish its size and disfigure its habit (Mathieu, Fl. Forest, ed. 4, 611). Its rapid growth, strong root-system, and its ability to thrive on poor sandy soil, have led to the employment of this species for the forestation of sand-dunes in France.

The tree can be recognized by its long stout leaves and persistent brown cones. Its leaf-section is peculiar in the remarkable size of the inner cells of the hypoderm, especially in the angles of the leaf.

Plate XXXII.

Figs. 275, 276, Cones. Fig. 277, Magnified leaf-section. Fig. 278, Magnified dermal tissues in the angle of the leaf.

PLATE XXXII. P. PINASTER (275-278), HALEPENSIS (279-283)

## 52. PINUS VIRGINIANA

- 1768 P. virginiana Miller, Gard. Dict. ed. 8.
- 1789 P. inops Aiton, Hort. Kew. iii. 367.

Spring-shoots multinodal, pruinose; branchlets pliant and tough. Bark-formation slow, the cortex not rifted for some years. Leaves binate, from 4 to 8 cm. long; resin-ducts medial, or with an occasional internal duct; hypoderm biform. Conelets with long tapering sharp scales. Cones from 4 to 6 cm. long, ovate or oblong-ovate, symmetrical, persistent, dehiscent at maturity; apophyses lustrous nut-brown, somewhat elevated along a transverse keel, the umbo salient, forming a long slender prickle with a broad base.

Western Long Island to central Georgia and north Alabama, and from eastern Tennessee to southern Indiana and southeastern Ohio. It is a low bushy tree in the north, but in the south and west it attains small timber-size and is locally exploited. It is hardy beyond the limits of its natural range, growing readily in the vicinity of Boston. Its short binate leaves, the persistent long prickles of its cone, and its tough branches, combine to distinguish this Pine from its associates. The obvious relationship of P. virginiana and P. clausa places the former in this, rather than in the preceding group.

Plate XXXIII.

Fig. 284, Cones. Fig. 285, Conelet and its enlarged spinose scale. Fig. 286, Leaf-fascicle, magnified leaf-section and more magnified dermal tissues of the leaf. Fig. 287, Buds.

### 53. PINUS CLAUSA

- 1884 P. clausa Vasey ex Sargent, Rep. 10th Cens. U. S. ix. 199.

Spring-shoots multinodal. Bark-formation slow, as in the preceding species. Leaves binate, from 5 to 9 cm. long; resin-ducts medial, or with an occasional internal duct, hypoderm biform when of two rows of cells. Conelets with long tapering acute scales. Cones from 5 to 8 cm. long, reflexed, ovate-conic, symmetrical, persistent, often serotinous; apophyses lustrous nut-brown, elevated along a transverse keel, the umbo forming a triangular persistent spine.

A species of limited range, confined to the sandy coast of Alabama and to Florida. It sometimes attains timber-size, but is usually a low spreading tree of no commercial importance and never seen in cultivation. It is recognized by its smooth branches, binate leaves and numerous, often multiserial, clusters of persistent, often closed, cones. It is associated with P. caribaea and, in the northern part of its range, it grows with the other Southern species. By its close resemblance it may be considered the serotinous form of P. virginiana.

Plate XXXIII.

Fig. 288, Three nodal groups of cones of the same year. Fig. 289, Conelet and its enlarged scale. Fig. 290, Leaf-fascicle and magnified leaf-section. Fig. 291, Larger form of the tree.

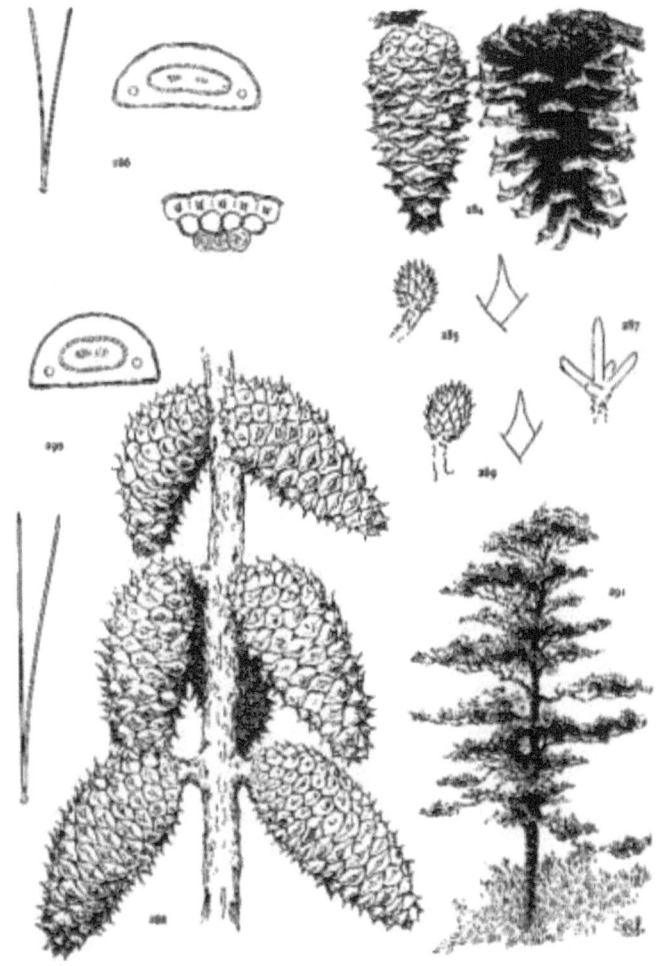

PLATE XXXIII. P. VIRGINIANA (284-287), CLAUSA (288-291)

## 54. PINUS RIGIDA

- 1768 P. rigida Miller, Gard. Dict. ed. 8.
- 1909 P. serotina Long, in Bartonia, ii. 17 (not Michaux).

Spring-shoots multinodal. Leaves ternate, from 7 to 14 cm. long; resin-ducts medial, or with an occasional internal duct, hypoderm biform. Scales of the conelet abruptly prolonged into a spine. Cones from 3 to 7 cm. long, ovate-conic, symmetrical, persistent, dehiscent at maturity or rarely serotinous; apophyses lustrous tawny yellow, elevated along a transverse keel, the umbo salient and forming the broad base of a slender sharp prickle.

A tree with bright green foliage in spreading tufts. The northern limit of its range is in southwestern New Brunswick, southern Maine, central New Hampshire and Vermont, the Thousand Islands of the St. Lawrence River and central Ohio. It ranges into Pennsylvania and Delaware at low levels and thence over the Alleghanies into northern Georgia. It is associated with P. strobus and P. resinosa and, further south, with P. virginiana. The cones are rarely serotinous, but it is remarkably like P. serotina in many characters, and is therefore placed in this group.

Plate XXXIV.

Fig. 292, Cones. Fig. 293, Leaf-fascicle, magnified section through a fascicle, and magnified dermal tissues of the leaf. Fig. 294, Upper part of a tree.

## 55. PINUS SEROTINA

- 1803 P. serotina Michaux, Fl. Bor. Am. ii. 205.

Spring-shoots multinodal. Leaves ternate, from 12 to 20 cm. long; resin-ducts medial or medial and internal, hypoderm biform. Conelet long-mucronate. Cones from 5 to 7 cm. long, subglobose or short-ovate, symmetrical, persistent, serotinous; apophyses lustrous tawny yellow, slightly elevated along a transverse keel, the umbo forming the broad base of a slender, rather fragile prickle.

This species is confined to low wet lands from southeastern Virginia to northern Florida and central Alabama. It is one of the associated six timber-Pines of the Southern States and the only one of them with serotinous cones. Its wood is of like value with that of P. taeda, the two species being constantly confused by lumbermen. It is never associated with P. rigida, but its resemblance to that Pine is so great that it may be regarded as its serotinous form. Its leaf is longer, its cone usually more orbicular and the prickle weaker.

Plate XXXIV.

Fig. 295, Cone. Fig. 296, Conelet and its enlarged scale. Fig. 297, Leaf-fascicle and magnified leaf-section.

84

### 56. PINUS PUNGENS

- 1803 P. taeda Lambert, Gen. Pin. i. t. 16, (as to cone). (not Linnaeus).
- 1806 P. pungens Lambert in Ann. Bot. ii. 198.
- 1852 P. montana Noll, Bot. Class Book, 340. (not Miller).

Spring-shoots multinodal. Leaves binate or ternate, from 3 to 7 cm. long; resin-ducts medial, or with an occasional internal duct, hypoderm biform. Scales of the conelet much prolonged into a very acute triangle. Cones from 5 to 9 cm. long, symmetrical or subsymmetrical, tenaciously persistent, serotinous; apophyses lustrous or sublustrous fulvous brown, much elevated along a transverse keel, the umbo forming a stout formidable spine, uniform or nearly uniform on all faces of the cone.

A mountain species ranging from central Pennsylvania to northern Georgia, with isolated stations in western New Jersey and Maryland. It is remarkable among the Pines of eastern North America for the size and strength of the spines of its cone. The armature resembles that of the cone of the western P. muricata, but with the difference that the western cone is strongly oblique, the anterior and posterior spines varying greatly in size.

Plate XXXIV.

Fig. 298, Cone. Fig. 299, Conelet and its enlarged scale. Fig. 300, Leaf-fascicle and magnified leaf-section.

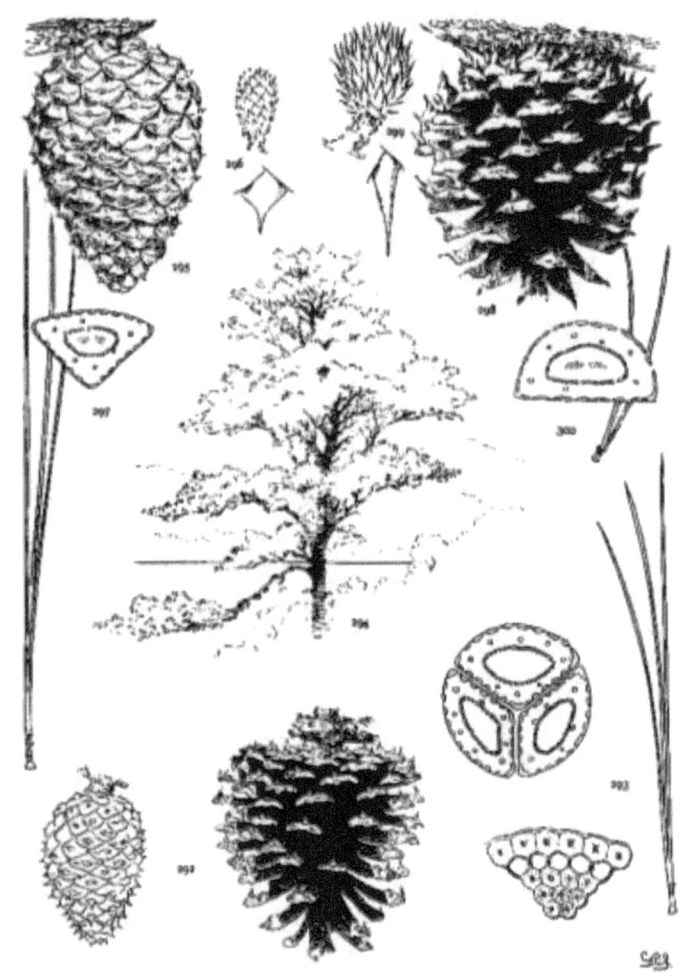

PLATE XXXIV. P. RIGIDA (292-294), SEROTINA (295-297), PUNGENS (298-300)

## 57. PINUS BANKSIANA

- 1803 P. Banksiana Lambert, Gen. Pin. i. 7. t. 3.
- 1804 P. hudsonia Poiret in Lamarck, Encycl. Méth. v. 339.

- 1810 P. rupestris Michaux f. Hist. Arbr. Am. i. 49, t. 2.
- 1811 P. divaricata Dumont de Courset, Bot. Cult. ed. 2, vi. 457.

Spring-shoots multinodal. Leaves binate, from 2 to 4 cm. long; resin-ducts medial, hypoderm biform. Conelets minutely mucronate. Cones from 3 to 5 cm. long, erect, ovate-conic, oblique, much curved or variously warped from the irregular development of the scales, serotinous; apophyses lustrous tawny yellow, concave, flat or convex, the umbo small and unarmed.

The most northern American Pine, growing near the Arctic Circle in the valley of the Mackenzie River, whence it ranges southeasterly to central Minnesota and the south shore of Lake Michigan, and easterly through the Dominion of Canada to northern Vermont, southern Maine, and Nova Scotia. In the northern part of its range it is the only Pine, but further south it is associated with P. strobus and P. resinosa. It is easily identified by its curious curved or deformed cones.

Plate XXXV.

Fig. 301, Cones. Fig. 302, Biserial cones of the same year. Fig. 303, Leaf-fascicle and magnified leaf-section. Fig. 304, Habit of the tree.

## 58. PINUS CONTORTA

- 1833 P. inops Bongard in Mém. Acad. Sci. St. Pétersb. ii. 163, (not Aiton).
- 1838 P. contorta Douglas ex Loudon, Arb. Brit. iv. 2292, f. 2211.
- 1853 P. Murrayana Balfour in Bot. Exp. Oregon, 2, f.
- 1854 P. Boursieri Carrière in Rev. Hort. 225, ff. 16, 17.
- 1868 P. Bolanderi Parlatore in DC. Prodr. xvi-2, 379.
- 1869 P. tamrac Murray in Gard. Chron. 191, ff. 1-9.
- 1898 P. tenuis Lemmon in Erythea, vi. 77.

Spring-shoots multinodal. Leaves binate, from 3 to 5 cm. long; resin-ducts medial, hypoderm biform. Conelets long-mucronate.

Cones from 2 to 5 cm. long, sessile, ovate-conic, symmetrical or very oblique, persistent, serotinous; apophyses lustrous tawny-yellow, flat or protuberant, on oblique cones abruptly larger on the posterior face; the umbo armed with a slender fragile prickle.

86

It grows from the valley of the Yukon, near the Alaskan boundary, along the Pacific coast to Mendocino county, California. It covers the plains and slopes of British Columbia and follows the Rocky Mountains into western Colorado, with an outlying station on the Black Hills of South Dakota. It grows on the Sierras and mountains of southern California and in northern Lower California. On the seashore this Pine is of low dense growth, but inland it is a slender tree with a long tapering stem. It is easily recognized by its very short leaves and very small cone.

Plate XXXV.

Fig. 305, Cones. Fig. 306, Leaf-fascicle and magnified leaf-section.

85

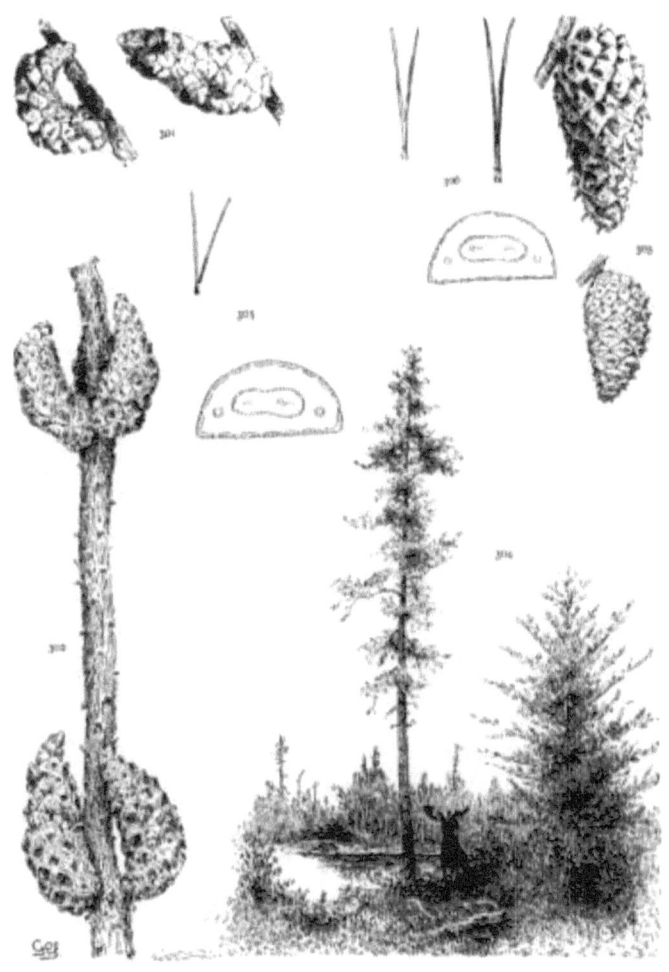

PLATE XXXV. P. BANKSIANA (301-304), CONTORTA (305, 306)

## 59. PINUS GREGGII

- 1868 P. Greggii Engelmann ex Parlatore in DC. Prodr. xvi-2, 396.

Spring-shoots uninodal and multinodal, pruinose. Bark-formation late, the branches and upper trunk smooth. Leaves ternate, from 7 to 10 cm. long, erect; resin-ducts medial, hypoderm of uniform thin-walled cells. Conelets mucronate. Cones from 6 to 12 cm. long, ovate-conic, oblique, serotinous, reflexed; apophyses lustrous tawny yellow, convex, the posterior gradually larger and more prominent than the anterior scales, the umbo flat or depressed, the mucro deciduous.

This species is known, at present, from specimens collected in the vicinity of the city of Saltillo, in northeastern Mexico. Were it not for the difference of bark it might be considered to be a northern variety of P. patula with shorter erect leaves. With both species the long peduncle of the conelet becomes overgrown by the basal scales of the ripe cone, which appears to be sessile. With both, the cones are in crowded nodal clusters, reflexed against the branch. They are so much alike that earlier descriptions of P. patula included the smooth gray bark of P. Greggii. The first correct description of the scaly red bark of P. patula appeared in the second edition of Veitch's Manual of Conifers.

Plate XXXVI.

Fig. 311, Cone. Fig. 312, Conelet. Fig. 313, Leaf-fascicle and magnified leaf-section. Fig. 314, Branch showing erect leaves.

## 60. PINUS PATULA

- 1831 P. patula Schlechtendal & Chamisso in Linnaea, vi. 354.

Spring-shoots multinodal, more or less pruinose. Bark-formation early, the scales deciduous, the upper trunk and branches red. Leaves prevalently ternate but sometimes in fascicles of 4 or 5, from 15 to 30 cm. long, slender and gracefully drooping; resin-ducts medial or with an occasional internal duct, hypoderm weak, of uniform thin-walled cells. Conelets mucronate. Cones from 6 to 11 cm. long, in crowded verticillate clusters, sessile, reflexed, ovate-conic, oblique, persistent and serotinous; apophyses lustrous nut-brown, more or less tumid, the posterior gradually larger than the anterior scales, the umbo flat or depressed, the mucro wanting.

Patula grows in the warm-temperate climates of Hidalgo, Puebla and Vera Cruz, in eastern and central Mexico. It can be at once recognized by its slender drooping foliage, its persistent cones, and its red upper trunk. It is cultivated in northern Italy and in the warmer parts of Great Britain.

Plate XXXVI.

Fig. 307, Cone. Fig. 308, Conelet. Fig. 309, Leaf-fascicle and magnified leaf-section. Fig. 310, Branchlet with drooping leaves.

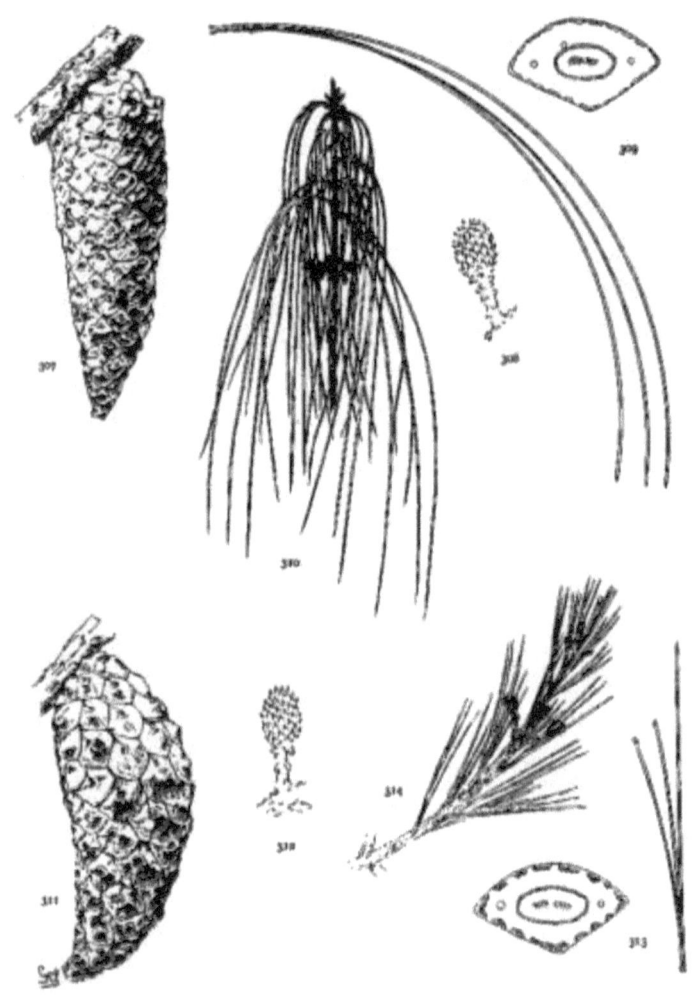

PLATE XXXVI. P. PATULA (307-310), GREGGII (311-314)

### 61. PINUS MURICATA

- 1837 P. muricata D. Don in Trans. Linn. Soc. xvii. 441.
- 1848 P. Edgariana Hartweg in Jour. Hort. Soc. Lond. iii. 217.

Spring-shoots multinodal. Leaves binate, from 10 to 15 cm. long; resin-ducts medial, hypoderm biform. Scales of the conelet prolonged into a triangular spine. Cones from 5 to 9 cm. long, in verticillate clusters, sessile, reflexed, ovate-conic, oblique, serotinous; apophyses lustrous nut-brown, 88 abruptly much larger on the posterior face of the cone, each armed with a formidable spine varying in size with the varying size of the apophysis.

This species grows on the coast of California, in scattered stations between Mendocino and San Luis Obispo Counties, and on the northwest coast of Lower California and on Cedros Island. It is recognized by its oblique cones, conspicuously spinose, indefinitely persistent and very serotinous. The unequal development of its cone-scales distinguishes the cone from the more symmetrically developed cone of P. pungens. Fruiting trees of P. muricata may be seen in the Royal Gardens at Kew.

Plate XXXVII.

Fig. 315, Cone. Fig. 316, Leaf-fascicle and magnified leaf-section.

### 62. PINUS ATTENUATA

- 1847 P. californica Hartweg in Jour. Hort. Soc. Lond. ii. 189, (not? P. californiana, Loiseleur).
- 1849 P. tuberculata Gordon in Jour. Hort. Soc. Lond. iv. 218, f. (not D. Don).
- 1892 P. attenuata Lemmon in Mining & Sci. Press, lxiv. 45.

Spring-shoots multinodal. Bark-formation late, the branches and upper trunk smooth. Leaves ternate, from 8 to 16 cm. long; resin-ducts medial or with one or more internal ducts, hypoderm biform. Scales of the conelet prolonged into a triangular spine. Cones from 8 to 16 cm. long, in verticillate clusters, sessile, reflexed, long-ovate, oblique, persistent and remarkably serotinous; apophyses lustrous tawny yellow, abruptly larger and more prominent on the posterior face of the cone, where they are usually prolonged into acute pyramids with a small incurved spine.

A tree of slender habit and gray-green foliage, the trunk studded with persistent nodal cone-clusters; growing on dry mountain slopes, from southwestern Oregon over the foot-hills of the northern mountains of California and its coastal ranges as far as the southern slopes of the San Bernardino Mountains. It attains its best development in the northern part of its range, but is never a tree of importance. The serotinous habit is more pronounced in this than in any other species. It is distinct from P. radiata, its nearest relative, by the color of the cone, by its smooth upper trunk and by its much smaller size.

The possibility of identifying P. californiana Loiseleur (Nouv. Duham. v. 293), through a cone said to have been sent to the Museum at Paris, may cause this name to be applied, by reason of its early date (1812), to some existing species. Don's radiata and tuberculata, although considered to be the same species, were nevertheless founded on different forms of the cone. Under a very narrow conception of specific limits tuberculata Don might therefore acquire specific rank. These considerations seem to make it advisable to abandon for this species the names californica Hartw. and tuberculata Gord. for the later name attenuata.

Plate XXXVII.

Fig. 317, Cone. Fig. 318, Magnified leaf-section.

### 63. PINUS RADIATA

- 1837 P. radiata D. Don in Trans. Linn. Soc. xvii. 442.
- 1837 P. tuberculata D. Don in Trans. Linn. Soc. xvii. 442.
- 1838 P. insignis Douglas ex Loudon, Arb. Brit. iv. 2265, f. 2171.
- 1841 P. Sinclairii Hooker & Arnott in Bot. Beechy Voy. 392, t. 93 (as to leaves).

Spring-shoots multinodal. Bark formation early, the branches and upper trunk rough. Leaves ternate or binate, from 10 to 15 cm. long; resin-ducts medial or with an occasional internal duct, hypoderm biform. Conelets mucronate, the mucro small and dorsal. Cones from 7 to 14 cm. long, in verticillate clusters, sessile, reflexed, ovate or oblong, oblique, serotinous; apophyses nut-brown, lustrous, tumid in various degrees, the posterior scales abruptly larger and very prominent, the umbo bearing the minute prickle or its remnant.

A tall tree with rich green foliage, growing on a strip of coast south of San Francisco, particularly in Monterey County. It grows also on the islands forming the Santa Barbara Channel and on the Island of Guadeloupe, Lower California. It is remarkably successful in the warmer climates of Europe and of Australasia. The species is distinct in its peculiar cone with rounded apophyses.

Plate XXXVII.

Figs. 319, 320, Cones. Fig. 321, Leaf-fascicle and magnified leaf-section. Fig. 322, Leaf-section from a binate fascicle. Fig. 323, Magnified dermal tissues of the leaf.

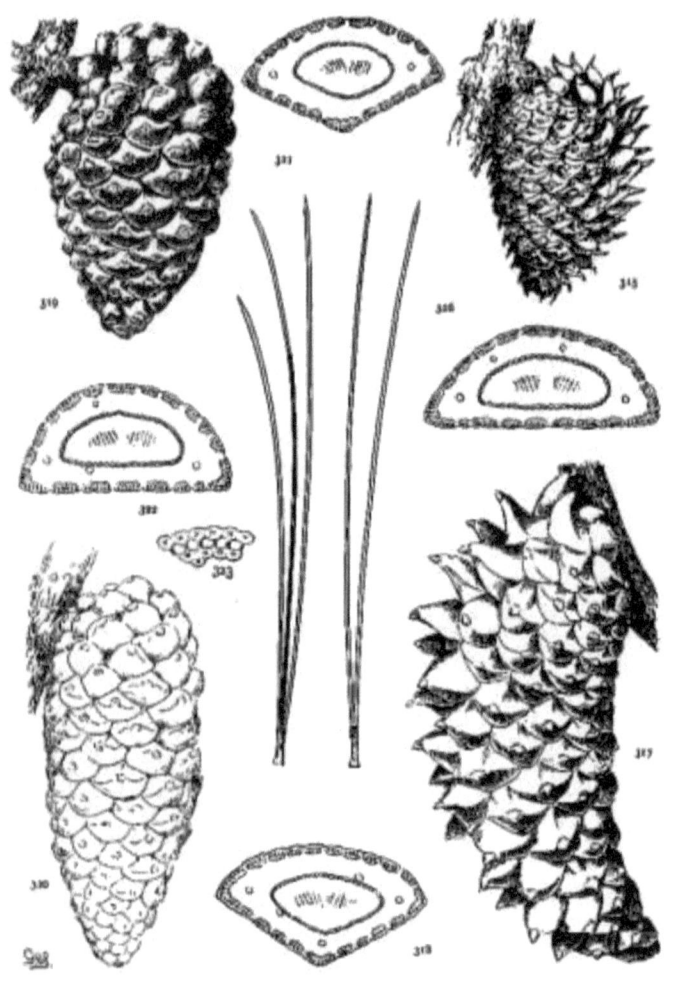

PLATE XXXVII. P. MURICATA (315, 316), ATTENUATA (317, 318), RADIATA (319-323)

## XIII. MACROCARPAE

Pits of the ray-cells small. Wing-blade of the seed thick. Cones large. Leaves long and stout.

This group is remarkable for the size of leaf, conelet, and cone. The peculiar thick seed-wing is more or less obscurely present among the species of the Insignes, but never attains the development that differentiates this group from all other Pines. The leaf-section is notable for the large amount of hypoderm and for the presence of both thick and thin outer walls of the endoderm-cells, both forms appearing in the same leaf.

| | |
|---|---|
| Wing-blade with a short membranous extension. | |
| Leaves in fascicles of 5 | 64. Torreyana |
| Leaves in fascicles of 3 | 65. Sabiniana |
| Wing-blade with a long membranous extension, leaves in fascicles of 3 | 66. Coulteri |

## 64. PINUS TORREYANA

- 1855 P. Torreyana Parry ex Carrière, Trait. Conif. 326.
- 1860 P. lophosperma Lindley in Gard. Chron. 46.

Spring-shoots uninodal, pruinose. Leaves in fascicles of 5, from 20 to 33 cm. long, very stout; resin-ducts medial, hypoderm uniform or somewhat multiform and of many cells. Conelets large, mucronate. Cones from 10 to 15 cm. long, on stout peduncles, broad-ovate, symmetrical, somewhat persistent; apophyses chocolate-brown, prominently pyramidal, the umbo salient and capped with a small mucro; seed-wing short, very thick, the dorsal surface of the nut spotted with the black remnants of the spermoderm.

A tree 10 or 12 metres high, often semi-prostrate in exposed positions, confined to a restricted area on the coast north of San Diego, California, and to the Island of Santa Rosa. This species resembles P. Sabiniana in the length of its seed-wing and in the color of its cone, but is distinct in the short triangular umbo, in its pentamerous leaf-fascicles and in the mottled dorsal surface of its nut.

Plate XXXVIII.

Fig. 324, Cone and seed. Fig. 325, Magnified leaf-section.

## 65. PINUS SABINIANA

- 1833 P. Sabiniana Douglas in Trans. Linn. Soc. xvi. 747.

Spring-shoots multinodal, pruinose. Leaves in fascicles of 3, from 20 to 30 cm. long; resin-ducts medial, hypoderm multiform. Conelets large, their scales tapering to a sharp point. Cones from 15 to 25 cm. long, reflexed, ovate, slightly oblique, persistent; apophyses chocolate-brown, very prominent, the curved umbo confluent with the apophysis and with it forming a very large talon-like armature with a sharp apex and a broad thick base; seed-wing very thick, with a short membranous margin, the dorsal surface of the nut uniform in color.

A tree with sparse gray-green foliage, growing in small groves on the foot-hills of the Sierra Nevada and Coast Ranges of California. Its three leaves and the uniform color of the nut distinguish it from 93 P. Torreyana. From P. Coulteri it differs in the length of the membranous portion of the seed-wing and in its gray-green leaves.

Plate XXXVIII.

Fig. 326, Cone. Fig. 327, Seed, nut and wing. Fig. 328, Magnified leaf-section.

PLATE XXXVIII. P. TORREYANA (324, 325), SABINIANA (326-328)

## 66. PINUS COULTERI

- 1837 P. Coulteri D. Don in Trans. Linn. Soc. xvii. 440.
- 1840 P. macrocarpa Lindley in Bot. Reg. xxvi. Misc. 62.

Spring-shoots multinodal, pruinose. Leaves in fascicles of 3, from 15 to 30 cm. long, very stout; resin-ducts medial, or with an occasional internal duct, hypoderm multiform and of many cells. Conelet very large, the scales tapering to a long sharp point. Cones from 25 to 35 cm. long, reflexed, ovate or oblong-ovate, somewhat oblique, persistent; apophyses sublustrous tawny yellow, very protuberant, with a narrow shoulder from which springs the umbo in the form of a large stout curved talon; seed-wing nearly equally divided between the very thick base and the membranous apex.

Remarkable among Pines for the size and weight of its cones, many times heavier than the longer cones of P. Lambertiana, illustrating the great change that the cone-tissues undergo in the gradual evolution of the species. It is a tree with dark-green foliage, growing from northern Lower California over the mountains of southern California to the Santa Lucia range and to Mt. Diablo. It is of no value except for fuel and for its large nuts. It is best recognized by its seed. The cone differs from the others of this group in its yellow color, not unlike that of boxwood.

Plate XXXIX.

Fig. 329, Cone of small size. Fig. 330, Seed, nut and wing. Fig. 331, Magnified leaf-section. Fig. 332, Conelet.

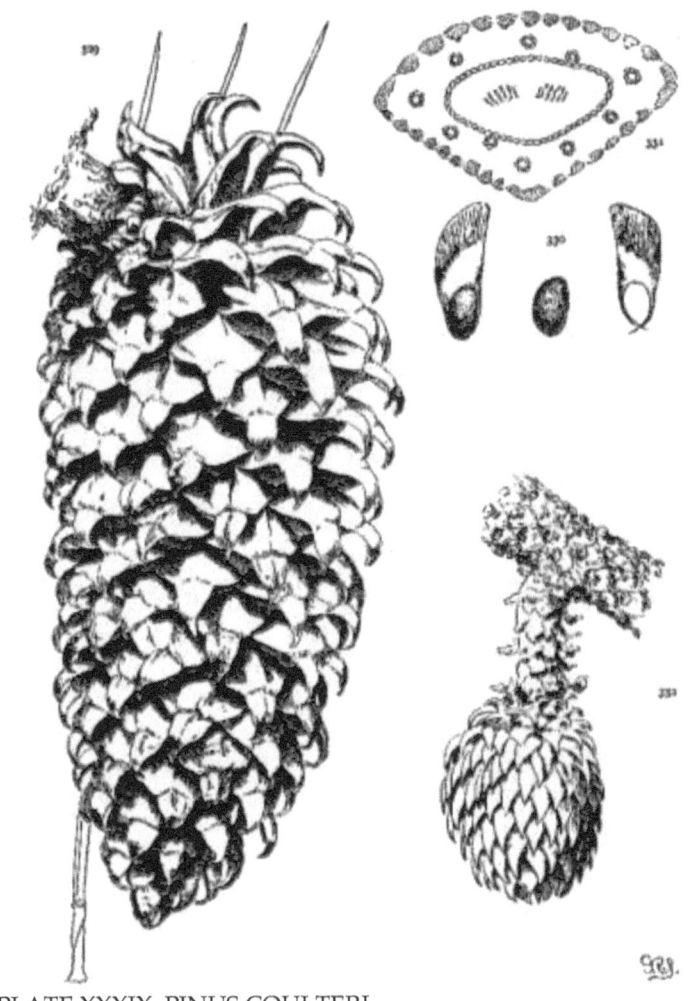

PLATE XXXIX. PINUS COULTERI

# INDEX

- Adnate wing, 16
- Apinus, Neck. — Pinus, 24
- Apophysis of cone, 10
- Armature of conelet, 7
- Articulate wing, 16
- Australes-Group, 62

- Balfourianae-Group, 42
- Bark, 18
- Bases of bracts decurrent and non-decurrent, 1
- Bast-tissue of cone, 14
- Biform hypoderm, 6
- Bloom on branchlet, 2
- Bracts, bases of, 1
- Branchlet, multinodal, 2
    - uninodal, 2
- Bud, latent, 2
    - leaf, 1
    - staminate, 1

- Caryopitys, Small = Pinus, 24
- Caryopitys edulis Small = Pinus cembroides, 38
- Cembra, Opiz = Pinus, 24
- Cembra-Subsection, 26
- Cembrae-Group, 26
- Cembroides-Group, 38
- Characters of the genus, 1
- Classification of species, 22
- Color of cone, 8
- Cone, apophysis of, 10
    - bast tissues of, 14
    - color of, 8
    - dimensions of, 8
    - oblique, 10

- - peduncle of, 8
  - persistent, 8
  - phyllotaxis of, 12
  - serotinous, 14
  - symmetrical, 10
- Conelet, lateral, 7
  - pseudolateral, 7
  - subterminal, 7
- Conspectus of Sections &c., 25
- Connective of pollen-sacs, 7
- Cotyledons, 1

- Decurrent bases, 1
- Definite phyllotaxis, 12
- Dermal tissue of leaf, 4
- Dimensions of cone, 8
  - leaf, 4
- Diploxylon-Section, 44
- Dorsal umbo, 8

- Endoderm, 8
- Evolutional characters, 20
- External resin-ducts, 6

- Fibro-vascular bundle, 6
- Flexiles-Group, 28
- Flowers, lateral, 7
  - pistillate, 7
  - staminate, 7
  - subterminal, 7

- Generic characters, 18
- Genus, characters of, 1
- Geographical distribution, 24
- Gerardianae-Group, 40
- Green tissue of leaf, 6

- Haploxylon-Section, 26
- Hypoderm, biform, 6
    - multiform, 6
    - uniform, 6

- Indefinite phyllotaxis, 12
- Insignes-Group, 76
- Internal resin-ducts, 6

- Lariciones-Group, 51
- Latent buds, 2
- Lateral flowers, 7
    - conelet, 7
- Leaf, dermal tissues of, 4
    - dimensions of, 4
    - fibro-vascular bundle of, 6
    - green tissue of, 6
    - persistence of, 4
    - primary, 1
    - resin-ducts of, 6
    - secondary, 2
    - stelar tissues of, 6
    - stomata of, 4
    - tissues of, 4
- Leiophyllae-Group, 44
- Longifoliae-Group, 46

- Macrocarpae-Group, 90
- Medial resin-ducts, 6
- Multiform hypoderm, 6
- Multinodal branchlet, 2
    - spring-shoot, 2

- Non-decurrent bases, 1

- Oblique cone, 10

- Paracembra-Subsection, 36
- Parapinaster-Subsection, 44
- Peduncle of cone, 8
- Persistence of leaf, 4
- Persistent cone, 8
- Phyllotaxis, of cone, 12
  - definite, 12
  - indefinite, 12
- Pinaster-Subsection, 50
- Pineae-Group, 48
- Pinus, 24
- Pinus abasica Carr. = halepensis, 78
  - alba-canadensis Prov. = strobus, 36
  - albicaulis Engelm., 27
  - Altamirani Shaw = Lawsonii, 68
  - apacheca Lemm. = ponderosa, 66
  - apulcensis Lindl. = pseudostrobus, 62
  - arabica Sieb. = halepensis, 78
  - aristata Engelm., 44
  - arizonica Engelm. = ponderosa, 66
  - Armandi Franch., 30
  - armena Koch = sylvestris, 54
  - attenuata Lemm., 88
  - australis Michx. = palustris, 70
  - austriaca Höss = nigra, 58
  - ayacahuite Ehrenb., 30 95
  - bahamensis Grise. = caribaea, 70
  - Balfouriana Balf., 42
  - Balfouriana Wats. = aristata, 44
  - Banksiana Lamb., 84
  - Beardsleyi Murr. = ponderosa, 66
  - Benthamiana Hartw. = ponderosa, 66
  - Bolanderi Parl. = contorta, 84
  - Bonapartea Roezl = ayacahuite, 30
  - Boursieri Carr. = contorta, 84
  - brachyptera Engelm. = ponderosa, 66
  - brutia Ten. = halepensis, 78
  - Bungeana Zucc., 40

- californica Hartw. = attenuata, 88
- canaliculata Miq. = Massoniana, 52
- canariensis Smith, 48
- caribaea Mor., 70
- carica Don = halepensis, 78
- cembra L., 27
- cembra Thunb. = parviflora, 32
- cembroides Gord. = Pinceana, 38
- cembroides Newb. = albicaulis, 27
- cembroides Zucc., 38
- chihuahuana Engelm. = leiophylla, 44
- clausa Vasey, 80
- contorta Dougl., 84
- coronans Litv. = cembra, 27
- Coulteri D. Don, 93
- Craigana Murr. = ponderosa, 66
- cubensis Grise. = occidentalis, 70
- cubensis Sarg. = caribaea, 70
- dalmatica Vis. = nigra, 58
- deflexa Torr. = ponderosa, 66
- densata Mast. = sinensis, 60
- densiflora Sieb. & Zucc., 52
- Devoniana Lindl. = Montezumae, 64
- divaricata Dum. Cours. = Banksiana, 84
- Donnell-Smithii Mast. = Montezumae, 64
- echinata Mill., 74
- Edgariana Hartw. = muricata, 86
- edulis Engelm. = cembroides, 38
- Ehrenbergii Endl. = Montezumae, 64
- eldarica Medw. = halepensis, 78
- Elliottii Engelm. = caribaea, 70
- Engelmanni Carr. = ponderosa, 66
- escarena Riss. = pinaster, 80
- excelsa Hook. = peuce, 34
- excelsa Wall., 34
- filifolia Lindl. = Montezumae, 64
- Finlaysoniana Wall. = Merkusii, 58
- flexilis James, 28
- flexilis Balf. = albicaulis, 27

- formosana Hay. = parviflora, 32
- Fremontiana Endl. = cembroides, 38
- Frieseana Wich. = sylvestris, 54
- funebris Kom. = sinensis, 60
- Gerardiana Wall., 42
- glabra Walt., 72
- Gordoniana Hartw. = Montezumae, 64
- Greggii Engelm., 86
- Grenvilleae Gord. = Montezumae, 64
- Griffithii McClell. = excelsa, 34
- halepensis Bieb. = nigra, 58
- halepensis Mill., 78
- Hamiltonii Ten. = pinaster, 80
- Hartwegii Lindl. = Montezumae, 64
- Heldreichii Chr. = nigra, 58
- Henryi Mast. = sinensis, 60
- heterophylla Small = taeda, 72
- heterophylla Sudw. = caribaea, 70
- hispanica Cook = halepensis, 78
- hudsonia Poir. = Banksiana, 84
- humilis Link = sylvestris, 54
- inops Ait. = virginiana, 80
- inops Bong. = contorta, 84
- insignis Dougl. = radiata, 88
- insularis Endl., 60
- Jeffreyi Balf. = ponderosa, 66
- kasya Royle = insularis, 60
- khasiana Griff. = insularis, 60
- Kochiana Klotzsch = sylvestris, 54
- koraiensis Mast. = Armandi, 30
- koraiensis Sieb. & Zucc., 26
- Lambertiana Dougl., 32
- lapponica Mayr = sylvestris, 54
- laricio Poir. = nigra, 58
- laricio Savi = pinaster, 80
- latifolia Sarg. = ponderosa, 66
- latisquama Engelm. = Pinceana, 38
- latteri Mason = Merkusii, 58
- Lawsonii Roezl., 68

- leiophylla Schl. & Cham., 44
- Lemoniana Benth. = pinaster, 80
- leucodermis Ant. = nigra, 58
- leucosperma Max. = sinensis, 60
- Lindleyana Gord. = Montezumae, 64
- Llaveana Schiede = cembroides, 38
- Loiseleuriana Carr. = halepensis, 78
- longifolia Roxb., 46
- lophosperma Lindl. = Torreyana, 90
- Loudoniana Gord. = ayacahuite, 30
- luchuensis Mayr, 56
- Lumholtzii Rob. & Fern., 46
- lutea Walt. = taeda, 72
- macrocarpa Lindl. = Coulteri, 93
- macrophylla Engelm. = ponderosa, 66
- macrophylla Lindl. = Montezumae, 64
- maderiensis Ten. = pinea, 48
- mandschurica Laws. = cembra, 27
- mandschurica Rupr. = koraiensis, 26
- maritima Ait. = nigra, 58
- maritima Lamb. = halepensis, 78
- maritima Poir. = pinaster, 80
- Massoniana Lamb., 52
- Massoniana Sieb. & Zucc. = Thunbergii, 56
- Mastersiana Hay. = Armandi, 30
- Mayriana Sudw. = ponderosa, 66
- Merkusii De Vriese, 58
- mitis Michx. = echinata, 74
- monophylla Torr. = cembroides, 38
- montana Lam. = cembra, 27
- montana Mill., 54
- montana Noll = pungens, 84
- Montezumae Lamb., 64
- monticola Dougl., 34
- morrisonicola Hay. = parviflora, 32
- mugho Poir. = montana, 54
- mughus Jacq. = sylvestris, 54
- mughus Scop. = montana, 54
- muricata D. Don, 86

- Murrayana Balf. = contorta, 84
- Nelsonii Shaw, 40
- nepalensis De Chamb. = excelsa, 34
- nigra Arnold, 58
- nigricans Host = nigra, 58
- nivea Booth = strobus, 36
- obliqua Saut. = montana, 54
- occidentalis H. B. K. = Montezumae, 64
- occidentalis Swartz, 70
- oocarpa Schiede, 78
- oocarpoides Lindl. = oocarpa, 78
- orizabae Gord. = pseudostrobus, 62
- osteosperma Engelm. = cembroides, 38
- Pallasiana Lamb. = nigra, 58
- palustris Miller, 70
- Parolinii Vis. = halepensis, 78
- Parryana Engelm. = cembroides, 38 96
- Parryana Gord. = ponderosa, 66
- parviflora Sieb. Zucc., 32
- patula Schl. & Cham., 86
- peninsularis Lemm. = ponderosa, 66
- pentaphylla Mayr = parviflora, 32
- persica Strangw. = halepensis, 78
- peuce Grise., 34
- pinaster Ait., 80
- pinaster Bess. = nigra, 58
- Pinceana Gord., 38
- pindica Form. = nigra, 58
- pinea Linn., 48
- pityusa Stev. = halepensis, 78
- ponderosa Dougl., 66
- pontica Koch = sylvestris, 54
- porphyrocarpa Laws. = monticola, 34
- Pringlei Shaw, 76
- prominens Mast. = sinensis, 60
- pseudostrobus Lindl., 62
- pumila Regel = cembra, 27
- pumilio Haenke = montana, 54
- pungens Lamb., 84

- pyrenaica David = halepensis, 78
- pyrenaica Lap. = nigra, 58
- quadrifolia Sudw. = cembroides, 38
- radiata D. Don, 88
- radiata Hook. & Arn. = Montezumae, 64
- recurvata Rowl. = caribaea, 70
- reflexa Engelm. = flexilis, 28
- resinosa Ait., 51
- resinosa Loise. = halepensis, 78
- resinosa Savi = sylvestris, 54
- rigida Mill., 82
- rotundata Link = montana, 54
- Roxburghii Sarg. = longifolia, 46
- Royleana Jam. = echinata, 74
- rubra Michx.= resinosa, 51
- rubra Mill. = sylvestris, 54
- rudis Endl. = Montezumae, 64
- rupestris Michx. = Banksiana, 84
- Russelliana Lindl. = Montezumae, 64
- Sabiniana Dougl., 90
- Salzmanni Dun. = nigra, 58
- sanguinea Lap. = montana, 54
- sativa Lam. = pinea, 48
- scipioniformis Mast. = Armandi, 30
- scopifera Miq. = densiflora, 52
- scopulorum Lemm. = ponderosa, 66
- serotina Long = rigida, 82
- serotina Michx., 82
- shasta Carr. = albicaulis, 27
- sibirica Mayr = cembra, 27
- Sinclairii Hook. & Arn. = Montezumae, 64
- = radiata, 88
- sinensis Lamb., 60
- squarrosa Walt. = echinata, 74
- strobiformis Engelm. = ayacahuite, 30
- strobiformis Sarg. = flexilis 28
- strobus Linn., 36
- strobus Thunb. = koraiensis, 26
- sylvestris Baumg. = nigra, 58

- sylvestris Gouan = halepensis, 78
- sylvestris Linn., 54
- sylvestris Lour. = Merkusii, 58
- sylvestris Mill. = pinaster, 80
- sylvestris Thunb. = Thunbergii, 56
- tabulaeformis Carr. = sinensis, 60
- taeda Blanco = insularis, 60
- taeda Lamb. = pungens, 84
- taeda Linn., 72
- tamrac Murr. = contorta, 84
- tatarica Mill. = sylvestris, 54
- tenuifolia Benth. = pseudostrobus, 62
- tenuis Lemm. = contorta, 84
- teocote Schl. & Cham., 68
- terthrocarpa Shaw = tropicalis, 52
- Thunbergii Parl., 56
- Torreyana Parry, 90
- tropicalis Mor., 52
- tuberculata D. Don = radiata, 88
- tuberculata Gord. = attenuata, 88
- uliginosa Neum. = montana, 54
- uncinata Ram. = montana, 54
- variabilis Lamb. = echinata, 74
- Veitchii Roezl = ayacahuite, 30
- virginiana Mill., 80
- Wilsonii Shaw = sinensis, 60
- Wincesteriana Gord. = Montezumae, 64
- Wrightii Engelm. = occidentalis, 70
- yunnanensis Franch. = sinensis, 60

- Pistillate flower, 7
- Primary leaf, 1
- Pseudolateral conelet, 7

- Resin-ducts of the leaf external, 6
    - internal, 6
    - medial, 6
    - septal, 6

- Secondary leaf, 2
- Sectional characters, 18
- Seed, wing of, 16
    - winged, 16
    - wingless, 16
- Septal resin-ducts, 6
- Serotinous cone, 14
- Species, classification of, 22
- Specific characters, 20
- Spring-shoot, 2
- Staminate flowers, 7
- Stelar tissues of leaf, 6
- Stomata of leaf, 4
- Strobi-Group, 30
- Strobus, Opiz = Pinus, 24
- Strobus strobus Small = Pinus strobus, 36
- Subsectional characters, 20
- Subterminal conelet, 7
    - flower, 7
- Symmetrical cone, 10

- Terminal umbo, 8
- Tissues of the cone, 12
    - leaf, 4
    - wood, 17

- Umbo of the cone dorsal, 8
    - terminal, 8
- Uninodal branchlet, 2
    - spring-shoot, 2
- Uniform hypoderm, 6

- Variation, 21

- Wing of seed, adnate, 16
    - articulate, 16

- Winged seed, 16
- Wingless seed, 16
- Wood, 17
- Wood-strands of the cone, 14
- Wood-tissues, 17

www.ingramcontent.com/pod-product-compliance
Lightning Source LLC
Chambersburg PA
CBHW031629210526
45464CB00004B/1809